Numerical Mathematics is a unique book that presents rudimentary numerical mathematics in conjunction with computational laboratory assignments. No knowledge of calculus or linear algebra is assumed, and thus the book is tailor-made for college freshmen and prospective mathematics teachers.

The material in the book emphasizes the algorithmic aspects of mathematics, making them viable through numerical assignments in which the traditional chalk-and-talk lecturer turns, in part, into a laboratory instructor. This book is not a numerical-methods book, containing ready-made computational recipes, but a guide for students to create algorithms for any assignment – in whichever programming language they use – on the basis of the underlying mathematics.

The computational assignments cover iterative processes, area approximations, solutions of linear systems, acceleration of series summation, interpolative approximations, and construction of computer library functions. Throughout, emphasis is placed on vital concepts such as error bounds, precision control, numerical efficiency, and computational complexity, as well as round-off errors and numerical stability.

NUMERICAL MATHEMATICS – A LABORATORY APPROACH

NUMERICAL MATHEMATICS – A LABORATORY APPROACH

SHLOMO BREUER and GIDEON ZWAS
Tel Aviv University

Published by the Press Syndicate of the University of Cambridge
The Pitt Building, Trumpington Street, Cambridge CB2 1RP
40 West 20th Street, New York, NY 10011-4211, USA
10 Stamford Road, Oakleigh, Melbourne 3166, Australia

First published 1993

Printed in the United States of America

Library of Congress Cataloging-in-Publication Data

Breuer, Shlomo.

Numerical mathematics: a laboratory approach / Shlomo Breuer, Gideon Zwas.

p. cm.

Includes index.

ISBN 0-521-44040-8

1. Numerical analysis. I. Zwas, Gideon. II. Title.
QA297.B68 1993

519.4 – dc20 92-36528
CIP

A catalog record for this book is available from the British Library.

ISBN 0-521-44040-8 hardback

Dedicated to the memory of
PROFESSOR SHLOMO BREUER
an outstanding researcher, lecturer, mathematics educator,
dear friend and colleague, and coauthor of this volume,
who unexpectedly passed away only a few weeks
after we completed the book.

G.Z.

Contents

Preface

Numerical Mathematics – A Laboratory Approach is a unique book that introduces the computational microcomputer laboratory as a vehicle for teaching algorithmic aspects of mathematics. This is achieved through a sequence of laboratory assignments, presupposing no previous knowledge of calculus or linear algebra, where the "chalk and talk" lecturer turns into a laboratory instructor. The computational assignments cover basic numerical topics that should be part of the mathematical education in the era of microcomputers.

In writing this book at the precalculus and pre–linear algebra level, we were mainly addressing an audience of four groups: first-year university students of mathematics, sciences, and engineering who have had no exposure to systematic calculus; students at teachers' training colleges who will be tomorrow's teachers of mathematics and computer science; superior high-school mathematics students; and scientific programmers at all levels. Various parts of this book were successfully tested on classes representative of each of these groups and subsequently modified. The material was received enthusiastically by high-school students who were members of Tel Aviv University's Math Club, some of whom are now faculty members of the School of Mathematical Sciences. The material was also welcomed by members of New York University's summer program for talented high-school students (held every summer at the Courant Institute of Mathematical Sciences and directed by Henry Mullish), and by several classes of in-service or future mathematics teachers at Tel Aviv University and at New York University.

An important feature of this book is that no previous knowledge of any specific programming language is assumed. Accordingly, the various algorithms are presented as a computer-language independent sequence of instructions in English. We have made a conscious effort to construct original elementary proofs of required results at the precalculus level, sacrificing complete generality at times but maintaining full rigor throughout.

Chapter 1 describes our philosophy of mathematical education, introduces our concept of the numerical laboratory, and details its educational potential. Chapters 2–8 are seven extensive computational assignments, comprising iterations for root extraction, area approximations, an algorithmic approach to linear systems, computations of π and e, convergence acceleration, interpolative approximation, and the construction of computer library functions. Important by-products of carrying out the assignments are various basic concepts, conspicuously absent from standard curricula, such as approximate solutions with error control, computational efficiency and complexity, the effects on solutions of small changes in the data, and the characterization of situations in which round-off errors are critical.

There are 20–30 exercises at the end of each chapter, with varying degrees of difficulty. A unique feature of reference to the exercises is employed, in that the reader is referred to each and every exercise at the appropriate location in the text, so that the means of its solution can be anticipated and considered. Solutions to selected exercises are given at the end of the book. Solutions to exercises that require extensive computer printouts have been omitted. Equations and inequalities are numbered sequentially in each chapter for easy referral and discussion.

The material in this book is based mainly on a series of ten articles published in *International Journal of Mathematics Education in Science and Technology* and in *Computers and Education* between 1983 and 1990 by the authors in collaboration with Dr. Judith Gal-Ezer of the Open University of Israel. A paper reviewing this series, entitled "Microcomputer laboratories in mathematics education," was published by the authors in *Computers and Mathematics with Applications*, Vol. 19, No. 3, pp. 13–34, 1990.

It is our pleasure to express our heartfelt thanks to Dr. Judith Gal-Ezer with whom we have collaborated for many years; to Prof. Eugene Isaacson of the Courant Institute of Mathematical Sciences for helpful discussions and many years of continuous support; to Prof. Marvin Marcus of the University of California at Santa Barbara for his encouragement; to Dr. Lynn Troyka of the City University of New York for her warm personal dedication and continuous help; to the many students and members of the high-school Math Club at Tel Aviv University for their productive feedback; to the members of New York University's summer program in computer science and its director Henry Mullish for their constructive suggestions; to our graduate students Ester Openheim, Ronit Hoffmann, Avital Stein, and Eugene Rodolphe for valuable additions; to scientific typists Miriam Hercberg, Gila Markowitz, and Joan Yichye-Shwachman for their careful typing and formatting of the manuscript; to the staff of Cambridge University Press and Dr. Alan Harvey in particular for their cooperation and understanding; and finally, with much affection, to our families, who stood behind us all these years and made it all worthwhile.

The mere presence of microcomputers in an educational institution, even where a programming language such as True Basic or Pascal is taught, in no way constitutes a new mode of teaching and learning. Using proper courseware, the full potential of microcomputers can be harnessed to improve the state of the art in education. Moreover, a new role can be played by mathematics teachers to achieve this objective, since their previous chalk-and-talk methods must henceforth be augmented by their active participation as *laboratory instructors*. Teaching mathematics using the assignments of this book, we believe, supplies the student and especially the future mathematics teacher with fertile ground for mathematical experiences that were not available before the personal computer era. These experiences will enhance and cultivate their mathematical intuition. Moreover, when exposed to these and similar assignments, students are molded in the spirit of numerical applied mathematics at an early stage, which is crucial to their entire mathematical point of view.

Tel Aviv University — Shlomo Breuer
May 1992 — Gideon Zwas

1
Mathematics in a numerical laboratory

1.1 The mathematical laboratory

Accumulated experience has shown that early emphasis on algorithmic thinking, augmented by actual computing, is indispensable in mathematical education. Recognizing the cardinal importance of the individual, active involvement of every student in the computational work (as opposed to mere demonstration by the teacher), we advocate the use of mathematical laboratories equipped with microcomputers. Optimally, a special room should be set aside for the mathematical laboratory. Failing that, physics or biology laboratories can be used since they tend to create the proper atmosphere. A pair of students is assigned to each microcomputer, as to a microscope in a biology laboratory, and spends a few hours a week working with the microcomputer in the laboratory.

The mere presence of an increasing number of microcomputers in various educational institutions, even those at which a programming language such as True-Basic or Pascal is taught, in no way constitutes a new mode of teaching and learning. The full potential of microcomputers and proper courseware should be harnessed to improve the state of the art in education. Moreover, a new role will be played by the mathematics teacher when traditional "chalk-and-talk" methods are augmented by active participation as a *laboratory instructor*. The numerous advantages of such computer-aided teaching of mathematics are detailed in Section 1.2.

The laboratory work will center around specific assignments, or modules, to be carried out by the participants at their own pace. Each of the following chapters constitutes one such module, containing various algorithms and the accompanying mathematical analysis. Frequently, we introduce new precalculus proofs to establish the required results. We assume throughout that the laboratory participants are familiar with the algorithm concept and with a programming language such as Pascal, which they will use to write and run their various programs.

We do not want to commit ourselves to any particular programming language, so we shall present the algorithms as in the following example, which is designed to find the sum of the square roots of the first n natural numbers (see also Exercises 1.1 and 1.2).

1. Input the value of n.
2. Set $k = 0$ and $s = 0$.
3. Replace the value of k by $(k + 1)$.
4. Replace the value of s by $(s + \sqrt{k})$.
5. If $k < n$, return to Step 3.
6. Output the value of s.
7. End.

As mentioned, computer-aided learning in the mathematical laboratory has many potential advantages, to which we turn our attention in the next section.

1.2 Potentials of laboratory learning

The following points bring out the educational potential of the mathematical laboratory:

Creativity coupled with delightful learning. In the laboratory, a higher percentage of the students are active (at their own pace) than under traditional learning circumstances. Creativity is stimulated by the gratifying dialogue with the computer. The student experiences the truth of the saying "mathematics is like kissing – the only way to discover its delights is by doing it."

Individual pace. The laboratory environment enables every pair of students to progress at a pace compatible with their ability (subject, of course, to some minimum goal required of everybody). The better pairs of students can tackle some of the "starred" assignments, without being held back by the traditional pace of an average class. "Double-starred" assignments are designed to challenge superior students to come up with original ideas, generalizations, improved algorithms, and so forth. Their progress should not be channeled toward the next assignment but toward a more profound mastery of the current subject. In this way, all students can realize their full potential.

Learning by discovery. The mathematical laboratory offers virtually unlimited opportunities for learning by discovery. The microcomputer enables the user to set up mathematical "experiments," test various conjectures, check nontrivial particular cases of a general proposition, and so forth. It goes without saying that not every student can take full advantage of this method of learning; but the better students can and will. Even in modern mathematical research, there are outstanding results whose origin can be traced to computational experimentation: The discovery of solitons is one case. Every student of mathematics should experience the gratification of personally discovering *some* mathematical rule. This method clearly enhances intuitive thinking, an essential component of the learning of mathematics.

Concretization of abstract ideas. Abstract mathematical concepts, such as the limit concept, can be made concrete and, thus, are likely to be vividly grasped and understood. Furthermore, the computational approach leads to an interplay between theoretical and numerical ideas, which undoubtedly improves the teaching process.

Getting down to earth. As pointed out by William E. Milne, "many know how to solve a problem but can't do it." The mathematical laboratory educates students not to fall into this category by training them to translate their theoretical ideas into algorithms that actually work. At the same time, the algorithmic thinking of students is cultivated – which, according to some researchers, should be placed at the center of mathematical education.

In-depth learning. While teaching a certain subject, we often find that we never fully understood certain subtle points; however, writing a computer program often shows us fundamental issues that we never paid attention to. Being a dummy (though a fast and powerful one), the computer carries out our instructions precisely but blindly. Any lack of in-depth understanding of a problem can lead to an imperfect program that will break down exactly when an unforeseen situation occurs (see Exercises 1.3 and 1.4).

No more tables. The laboratory is the most natural means of doing away with tables, which have traditionally been used by students as "black boxes" without the faintest understanding of their construction. We are not advocating the introduction of new black boxes by using, say, the logarithmic built-in function of the microcomputer; the student will learn what is behind such built-in functions as part of the material covered in the mathematical laboratory.

New vital aspects in the teaching of mathematics. The mathematical laboratory introduces and stresses, some vital concepts that are conspicuously absent from standard curricula – for example, approximate solutions with error control, computational efficiency and complexity, the influence of small changes in the data on the overall solution, characterization of situations in which the effect of round-off errors is critical, and methods for the construction of built-in, computer library functions.

Computational experimentation should be part of modern mathematical education. In fact, there has always been an experimental side to mathematics, as reflected in Leonhard Euler's saying that "the properties of numbers have usually been discovered by observation, well before their validity has been confirmed ... it is by observation that we increasingly discover new properties, which we next do our utmost to prove." Computer-aided learning of mathematics by means of the modules presented herein, supplies students with a fertile ground for mathematical experiences that were never available before the computer era. This enhances and cultivates their mathematical intuition and molds them in the spirit of numerical applied mathematics at an early stage – so crucial to their entire mathematical point of view.

1.3 Relative errors and significant digits

One of the central ideas in the modules is the solution of mathematical problems by various methods of approximation. It is natural, therefore, to inquire into the quality of the approximation employed or, alternatively, to estimate the error incurred. We shall usually be interested in approximation methods that enable us to make the error as small as we please, so that we can compute the solution to any desired accuracy. But what is actually meant by saying that the error is sufficiently small? Clearly, an error of half an inch is rather small when we are measuring the length of a jumbo jet, but the same error is disastrous if we are fitting a shoe to a customer's foot. Consequently, we are often compelled to test the *relative error*, which is the ratio of the absolute error to the approximation. Of course, when we use the term *error* we really mean an *error bound*, which must be derived in each given case. If we could find the exact error itself, we would also know the exact solution.

In this connection, the concept of significant figures should be clarified. In any given number, the significant figures are all the digits, counting from the first nonzero digit on the left. Thus, the number 0.00018 has two significant figures, the number −0.01503 has four, and the number 563.0081 has seven. If we represent these numbers in the form $1.8 \cdot 10^{-4}, -1.503 \cdot 10^{-2}$, and $5.630081 \cdot 10^2$, that is, in the form $m \cdot 10^k$ (where $1 \leq |m| < 10$ and k is an appropriate integer), then the significant figures are those of m (see Exercises 1.5 and 1.6). The preceding representation is called *scientific notation* and is widely used in pocket-calculators and microcomputers. A closely related concept in a computational context is the floating point, which is discussed in Section 1.4.

Now, suppose we want to construct a certain approximation method and use it to attain q correct significant figures in the result. Then we must require that

$$\left|\text{relative error}\right| < \frac{1}{2} \cdot 10^{-q} \tag{1.1}$$

In other words, we must ensure that the approximation coincides with the exact solution up to q significant figures whereas the $(q+1)$th figure may contain an error less than five. To see the significance of this, suppose the exact value of the required result is 0.00000462 and our approximation is 0.00000420. In this case, the absolute error is less than 10^{-6}, but our approximation possesses only one correct significant figure. Indeed, the relative error in this case is $(0.00000462 - 0.00000420)/0.00000420 = 0.10$ (an error of 10%), which is far from being less than 10^{-6}. Thus, we can obtain the desired number of significant figures only if the relative error is sufficiently small. But because we do not know the exact solution in general, we must replace the absolute error by an appropriate error bound when we estimate the relative error. The next section describes how these concepts tie in with actual numerical computations on the computer.

1.4 Floating point and round-off errors

Suppose we have a computing machine in which numbers are represented in scientific notation and arithmetical operations are carried out in accordance with the usual power laws. For example, if $a = 6.8 \cdot 10^4$ and $b = 9.7 \cdot 10^5$, then

$$a + b = 0.68 \cdot 10^5 + 9.7 \cdot 10^5 = 10.38 \cdot 10^5 = 1.038 \cdot 10^6 \tag{1.2}$$

$$a \cdot b = 6.8 \cdot 10^4 \cdot 9.7 \cdot 10^5 = 65.96 \cdot 10^9 = 6.596 \cdot 10^{10} \tag{1.3}$$

Note that the final answers are also represented in scientific notation $m \cdot 10^k$, where $1 \leq |m| < 10$; to achieve that, the decimal point is shifted (it floats) to its appropriate location. We say that our computing machine operates with *floating-point arithmetic* on a decimal basis (see Exercise 1.7).

Modern computers operate with floating-point arithmetic, but on a binary basis. Accordingly, the numbers are represented in the form $m \cdot 2^k$, where $1 \leq |m| < 2$ and k is an appropriate integer. The numbers k, m, and 2, of course, are themselves stored in the computer in binary form, using only zeros and ones. For example, 8 is represented as 1000, 0.25 as 0.01, and -5.5 as -101.1 (see Exercises 1.8 and 1.9). The use of floating-point arithmetic makes it possible to store very large numbers in the computer, such as the binary equivalent of $1.735 \cdot 10^{20}$, as well as very small numbers such as the binary equivalent of $5.8 \cdot 10^{-18}$.

The integer k and the number m $(1 \leq |m| < 2)$ are stored in the computer's memory, occupying a fixed "length" (number of binary digits, known as *bits*) that is typical of a given computer. The storage of m usually causes round-off errors, which we shall explain using decimal representations for clarity. For example, it is clear that $m = \frac{7}{9} = 0.777\ldots$ must be rounded off because it is impossible to store a number of infinite length; it is actually stored as $m = 0.77\ldots78$. Similarly, we encounter round-off errors as a result of adding a very small number to a very large one; for example, in a computer capable of handling 10 decimal places we have

$$
\begin{aligned}
7.68 \cdot 10^{12} + 2.34 \cdot 10^{2} &= (7.68 + 0.000000000234) \cdot 10^{12} \\
&= 7.680000000234 \cdot 10^{12} \qquad (1.4)
\end{aligned}
$$

but this result will be stored as $7.6800000002 \cdot 10^{12}$, and the round-off error is manifest. The situation resembles weighing an elephant on an appropriate scale, then weighing it again with a few flies on its back. Loss of accuracy due to the finite storage-length of m is also caused by the subtraction of two close numbers. For example,

$$
4.7584366213 - 4.7584366171 = 0.0000000042 = 4.2 \cdot 10^{-9} \qquad (1.5)
$$

and we see that although we started with two numbers having 11 correct significant figures, the result has only 2 correct significant figures. In fact, the final result will be stored as $4.2000000000 \cdot 10^{-9}$, where the nine zeros are meaningless. Such errors are referred to as *cancellation errors*. Clearly, all the loss-of-accuracy phenomena that results from the fixed storage-length are equally true in binary representation. If we had an ideal computer capable of storing numbers of infinite length, all the loss-of-accuracy problems would be eliminated. Because this is not the case, we must take these inaccuracies into account whenever we assess the quality of a computer-generated numerical result.

1.5 Generality of proofs

Mathematics, by its very nature, strives to discover and prove results and theorems with a maximal degree of generality. For example, in Euclidean geometry it is proved that the sum of the angles of *any* triangle is 180°; attention is not restricted, say, to isosceles triangles. The goal of complete generality usually entails a considerable intellectual effort, as well as longer proofs. Needless to say, a mathematical proof is acceptable only if it is a rigorous proof, a principle that must be emphasized throughout mathematical education. The attempt to present rigorous mathematical proofs for very general cases causes difficulties for precalculus (even cocalculus) classes, and therefore we must compromise somewhere. Because it is of cardinal importance, in our view, to preserve rigor, we advocate the occasional sacrifice of complete generality in the mathematical laboratory. For example, in some of the suggested modules we shall prove certain theorems about convex functions (convexity being defined geometrically), rather than about a wider class of, say, continuous functions. This simplifying assumption of convexity restricts the generality, which is the price we pay to be able to construct our own precalculus, rigorous proofs for the case at hand. The principle of occasional relaxation of generality, while maintaining rigor throughout, will guide our work on the various assignments in the mathematical laboratory.

Frequently we find that a certain restriction of generality is only apparent. For example, suppose we want to prove certain properties of functions that do not change sign in a given interval. Accordingly, we may assume, *without loss of generality*, that the underlying functions are nonnegative ($f(x) \geq 0$ for every x in the interval). We make use of the assumption $f(x) \geq 0$ throughout the proof, but this does not restrict the generality in any way because we can take $g(x) = -f(x)$ and repeat the proof with minor modifications. Thus, concepts such as *generality* and *loss of generality* must be part of mathematical education from its inception; the same goes for concepts such as *significant figures* and *computation to a desired degree of accuracy*.

1.6 Computational efficiency

Among the development of numerical methods for computing various approximations, we find two main categories: methods designed for permanent software, to be used a vast number of times, such as the built-in computer library functions ($\sqrt{x}$, $\sin x$, $\ln x$, etc.); and ad hoc methods, intended for auxiliary problems and for occasional application. For ad hoc methods, simplicity is preferable to preparatory mathematical sophistication (whose objective is computational efficiency); on the other hand, a numerical method designed for permanent use merits every conceivable preparatory effort to improve computational efficiency because the many small gains will accumulate.

Improvement of overall computational efficiency can result from two sources. An *improved numerical method* that is based on a worthwhile preparatory analysis may provide a faster approach to the solution (see Chapter 2), and a *rearrangement of the computational steps* of a given method may reduce computing time. The following is an example of the latter.

Suppose an assignment calls for the evaluation of the polynomial

$$y = P_n(x) = a_n x^n + a_{n-1} x^{n-1} + \cdots + a_1 x + a_0 \tag{1.6}$$

for many values of x, given the coefficients $a_0, a_1, \ldots, a_n$. If we evaluate y directly from (1.6), we need $(2n-1)$ multiplications and n additions for every value of x. On the other hand, if we rearrange (1.6) into Horner's form, which is,

$$y = P_n(x) = (\cdots((a_n x + a_{n-1})x + a_{n-2})x + \cdots + a_1)x + a_0 \tag{1.7}$$

only n multiplications and n additions are needed for the same job – clearly a more efficient procedure.

At this point, the following questions arise naturally: Is it possible to perform preliminary (one-time) preparations that will cast $P_n(x)$ into an even more efficient form? Is there a most efficient form? Such questions pertain to a subject called *computational complexity* (see Exercise 1.10). A central concern of computational complexity is to analyze problems of size n (e.g., evaluation of nth-degree polynomials, solution of n linear equations with n unknowns, a set of n inequalities) and to determine the total computational work as a function of n. In other words, is this computational work – that is, the complexity – proportional to $n!$, 2^n, n^6, n^2, $n \log n$, n, $\sqrt{n}$, or some other function of n? The complexity determines the total computing time and may occasionally render a problem practically insolvable (nontractable). This may be the case for $n = 100$, say, even when the largest and fastest modern computer is being used. Of great interest is the estimation of the intrinsic complexity of a given problem compared with that of the specific solution algorithm we are using. The greater the gap between the two, the greater the need to seek a more efficient algorithm.

Recently, parallel computers have appeared, which are capable of executing many operations simultaneously and thereby increasing computational efficiency immensely. This opens a new field of research – the development of algorithms whose parallel components are as large as possible.

1.7 Iterations – Stepwise approximations

Approximating the solution of a problem to any desired degree of accuracy, by a suitable approximation method, is an issue that is common to most of the mathematical laboratory assignments. Noteworthy is the process of stepwise approximation known as an *iterative method.* The idea is to start with a reasonable initial approximation reflecting the specific problem and to follow this with a corrective procedure, producing an improved approximation. Thus, one iteration has been completed. In the next iteration, the same corrective procedure is applied to the improved approximation, generating an even better one; and so on. The iterative method, then, consists of a set of consecutive calculations, each of which generates an approximate solution based on the approximation generated by its predecessor. Needless to say, we are interested in iterations that generate ever-improving approximations and that terminate as soon as some preassigned condition is met. The process is like parking a car in the space between two parked cars. The driver begins by moving the car to an initial approximate position and then moves forward and backward, correcting the car's position, until its wheels are sufficiently close to the curb (i.e., the preassigned condition is met). Other activities such as tuning a guitar, writing and debugging a computer program, composing and perfecting a song, or custom-tailoring a suit can also be viewed as iterations.

Regarding the criterion that tells us when to terminate an iterative process, it is often possible to obtain an a priori bound on the error incurred at any stage of the iterative process – that is, a bound on the deviation of the current approximation from the true solution. Thus, the iterations are terminated as soon as the error bound, and a fortiori, the error itself, is sufficiently small. In later chapters, laboratory participants will be guided toward making a considerable mathematical effort to determine such error bounds because this also makes it possible to predetermine the number of iterations required to achieve a prescribed accuracy.

As our first laboratory assignment, we shall take up iterations designed to compute square roots, cube roots, etc. Among other things, the accompanying analysis will shed light on the construction of the computer library square-root function. We shall turn to this subject in the next chapter.

Exercises

1.1 Write a verbal algorithm (a sequence of instructions in English) that finds $n!$ for a given integer n.

1.2 Write a verbal algorithm that finds the sum $(1 + 1/1! + 1/2! + \cdots + 1/n!)$ for a given integer n.

1.3 Write the solutions of the quadratic equation $ax^2 + bx + c = 0$, where $a > 0$. Find what happens to the solution when the value of a is gradually decreased and made to approach zero. (Start with $b = -2$, $c = -1$, and a taking on the sequence of values $0.1, 0.01, 0.001, \ldots$.)

1.4 Given the quadratic equation $10^{-q} \cdot x^2 - (1.4 + 2.5 \cdot 10^{-q})x + 3.5 = 0$, whose left-hand side factors into $10^{-q}(x - 2.5)(x - 1.4 \cdot 10^q)$. Clearly, the exact solutions are $x_1 = 1.4 \cdot 10^q$ and $x_2 = 2.5$. Use the standard formula to compute the solutions on a computer or a calculator, for $q = 12, 11, 10, \ldots, 2$. Note that you will obtain the correct value of x_1 whereas the value of x_2 will be way off for the larger values of q. Explain why this is so. Now compute the solutions, using

$$x_1 = \frac{-b + s\sqrt{b^2 - 4ac}}{2a} \quad \text{where} \quad s = \begin{cases} 1, & b \leq 0 \\ -1, & b > 0 \end{cases}$$

$$x_2 = \frac{c}{ax_1} \quad \text{since} \quad x_1 x_2 = \frac{c}{a}$$

and explain the disappearance of the error in the value previously obtained for x_2. This demonstrates the need for careful analysis, taking into account irregular input, even when one writes a program for as simple a problem as the solution of a general quadratic equation.

1.5 How many significant figures are there in the following numbers: 1703.05, −63.004, 0.007003, −101.01, 17.300, 0.000001433, 6,201,304.1, and 3.48.

1.6 Write the numbers of Exercise 1.5 in scientific notation.

1.7 Show that the representation of a given number in scientific notation is unique. (Assume two distinct representations, and deduce a contradiction.)

1.8 Write the following decimal numbers in binary form: 2, 0.5, 16, 7.25, 21, 0.125, and 0.1.

1.9 The following numbers are written in binary form: 101.101, 11111000100, 11.11, 0.001, 111, 1001, and $0.00110011\overline{00}\overline{1100}\ldots$, which is a binary periodic fraction whose period is 1100. Write those numbers in decimal form.

1.10 The evaluation of $P_4(x) = a_4x^4 + a_3x^3 + a_2x^2 + a_1x + a_0$, for a given value of x, requires seven multiplications and four additions, compared with four multiplications and four additions for Horner's form $P_4(x) = (((a_4x + a_3)x + a_2)x + a_1)x + a_0$. Show that $P_4(x)$ may be cast into the form $P_4(x) = a_4[x(x + \alpha) + \beta][x(x + \alpha) + x + \gamma] + \delta$, and express α, β, γ , and δ in terms of a_0, a_1, a_2, a_3, and a_4. Note that $x(x + \alpha)$ must be computed only once. The evaluation of α, β, γ, and δ is carried out, of course, only once, as a preparatory step.

Remark: The computationally economical form just shown was generalized to an nth-degree polynomial by Donald Knuth in 1962; it requires only about $(n/2 + 1)$ multiplications and $(n + 1)$ additions. For large n, the computational efficiency is manifest.

2
Iterations for root extraction

2.1 Introduction

As our first numerical laboratory assignment we present iterations that are designed to compute square roots, cube roots, and so on, of a given positive number. In Chapter 1 we have become familiar with the general idea of an iterative process, whose purpose is the successive approximation of a well-defined numerical target. We adopt a computer library approach throughout, that is, construction of permanent software that ought to be both efficient and fully automatic. In our case, this means that for a numerical input $d > 0$, the output $\sqrt[k]{d}$ (for integral $k \geq 2$) will be generated rapidly, with prescribed accuracy, and without any external interference.

2.2 Range reduction

It is expedient to start the root-finding process with an appropriate preparatory procedure known as *range reduction.*

Henceforth, we take d to be any positive number. Limiting ourselves to positive numbers is acceptable because we can easily dispose of the cases $d \leq 0$. For $d = 0$, we obtain the trivial output $\sqrt[k]{d} = 0$ whereas for $d < 0$ we have two cases. For k odd we proceed according to the identity $\sqrt[k]{d} = -\sqrt[k]{-d}$ whereas for k even the algorithm outputs the message "no real root in this case." This is an appropriate time for the laboratory instructor to incorporate a brief discussion of complex numbers, as supplementary material or as review of a subject previously studied.

A positive number d is represented in the computer in the form

$$d = m \cdot 2^i \tag{2.1}$$

where $1 \le m < 2$ and i is an appropriate integer. Using (2.1), we can also represent d in the form

$$d = s \cdot 4^j \qquad \begin{cases} s = m, \quad j = \dfrac{i}{2}, & \text{for } i \text{ even} \\ s = 2m, \quad j = \dfrac{(i-1)}{2}, & \text{for } i \text{ odd} \end{cases} \tag{2.2}$$

in which, of course, $1 \le s < 4$. The representation (2.2) is especially adapted to the computation of square roots ($k = 2$), and it yields

$$\sqrt{d} = \sqrt{s} \cdot 2^j \tag{2.3}$$

Here, $1 \le \sqrt{s} < 2$. For example, $\sqrt{384} = \sqrt{1.5 \cdot 2^8} = \sqrt{1.5 \cdot 4^4} = \sqrt{1.5} \cdot 2^4$, with $d = 384$, $m = 1.5$, $i = 8$, $s = 1.5$, $j = 4$. Again, $\sqrt{224} = \sqrt{1.75 \cdot 2^7} = \sqrt{3.5 \cdot 4^3} = \sqrt{3.5} \cdot 2^3$, with $d = 224$, $m = 1.75$, $i = 7$, $s = 3.5$, $j = 3$. Both examples confirm that indeed $1 \le \sqrt{s} < 2$.

This analysis shows that to compute $\sqrt{d}$ for a given $d > 0$, it is sufficient to find $\sqrt{s}$ for the corresponding s, in accordance with (2.2). Thus, without loss of generality, we henceforth concentrate our efforts on the construction of an iterative algorithm for the computation of $\sqrt{s}$, where s is limited to the reduced range $1 \le s < 4$. Note that the multiplications indicated on the right-hand sides of (2.2) and (2.3) are just *binary shifts*, which consume very little computer time – indeed, less time than addition.

At this point we suggest that the laboratory participants construct an analogous preparatory procedure for the computation of $\sqrt[3]{d}$. The reduced range is different in this case, but the cube root also satisfies $1 \le \sqrt[3]{s} < 2$. In this connection see Section 2.7 and Exercises 2.1 and 2.2.

Because the required root always lies in the interval $[1, 2)$, the absolute error $|\sqrt{s} - \text{approximation}|$ and the relative error $|\sqrt{s} - \text{approximation}|/\sqrt{s}$ are about equal. Consequently, if these two errors do not exceed $\frac{1}{2} \cdot 10^{-q}$, neither will the relative error associated with the computation of $\sqrt{d}$. All we have to do is to multiply both numerator and denominator of the quantity $|\sqrt{s} - \text{approximation}|/\sqrt{s}$ by the factor 2^j, appearing in (2.3). The same considerations apply to $\sqrt[k]{d}$, $k \ge 2$ (see Exercises 2.1 and 2.2).

2.3 Bisection method for square roots

Having finished the preparatory range-reduction procedure, we are now ready for actual root-finding iterations. Many authors suggest the use of the so-called bisection method for the iterative solution of $f(x) = 0$, in general. We shall begin with this method, but mainly as a point of

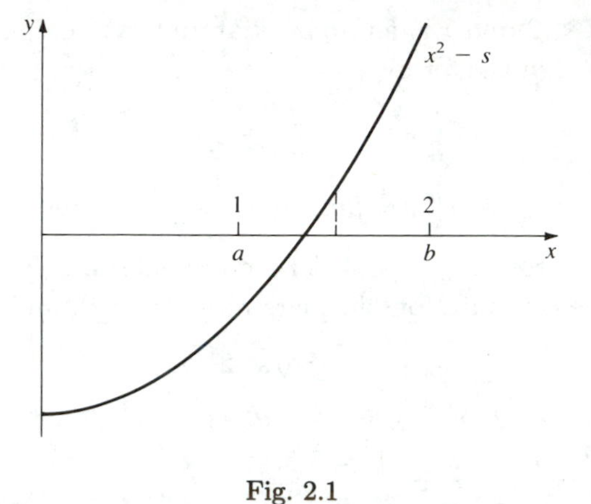

Fig. 2.1

departure toward developing more efficient methods and establishing a yardstick for comparisons.

Consider the graph in Fig. 2.1 of $y = f(x) = x^2 - s$ over the relevant interval $[1, 2]$, where $y(1) = 1 - s \leq 0$ and $y(2) = 4 - s > 0$. Thus, except for the special case $s = \sqrt{s} = 1$, the required value $\sqrt{s}$ lies between 1 and 2, precisely where the graph of the parabola intersects the x-axis. We denote by $a = 1$ and $b = 2$ the endpoints of our initial interval. Now let $x = (a + b)/2$ (i.e., the midpoint of the interval), and compute $y = x^2 - s$. If $y < 0$, we take x as our new a, and if $y > 0$, we take x as our new b (if $y = 0$, we have found the exact value of $\sqrt{s}$, and the computation terminates). In any case, we now have a new interval $[a, b]$, whose length is half its predecessor's, with $y(a) < 0$ and $y(b) > 0$. Thus, the *bisection* of the initial interval yielded a new interval that contains the root. Continuing in this fashion iteratively, we eventually obtain an interval $[a, b]$ whose length does not exceed $\frac{1}{2} \cdot 10^{-q}$, which contains $\sqrt{s}$. Thus, the bisection algorithm zeros-in on the required value $\sqrt{s}$ and can be formulated in the following way:

1. Input the value of q (desired number of correct figures).
2. Set $a = 1$, $b = 2$, and $E = \frac{1}{2} \cdot 10^{-q}$.
3. Input the value of s ($1 \leq s < 4$, after range reduction).
4. Compute $x = (a + b)/2$ and $y = x^2 - s$.
5. If $y = 0$, go to Step 8.
6. If $y > 0$, then $b = x$, else $a = x$.
7. If $(b - a) > E$, return to Step 4.
8. Output the values of x and y.
9. End.

Had we interchanged Steps 7 and 8, the results of each iteration would have been printed out, demonstrating the stepwise approach to $\sqrt{s}$. The instructor should point out that this approach to $\sqrt{s}$ is not necessarily one-sided. Thus, the deviation of the approximation from the true value of $\sqrt{s}$ does not necessarily decrease every time we iterate. Eventually, however, it will become as small as we please (see Exercise 2.3).

The bisection method hinges on the properties $y(a) < 0$, $y(b) > 0$ (or else $y(a) > 0$, $y(b) < 0$) and on the fact that the graph intersects the x-axis exactly once in $[a, b]$. Consequently, laboratory participants should be able to construct algorithms for the solution, to any desired accuracy, of complicated equations such as $x^7 + 2x - 200 = 0$ or $2^x + x - 5 = 0$. In this connection see Exercises 2.4 and 2.5.

At this point a natural question presents itself. If bisection is guaranteed to zero-in on the required root, why not use trisection? Wouldn't trisection work faster because each iteration would cut the interval into three thirds rather than two halves? It turns out that such is not the case. In bisection, each iteration requires one evaluation of $f(x)$, whereas trisection requires one evaluation per iteration in about one-third of the iterations and two evaluations per iteration in the remaining two-thirds. It follows that trisection requires an average of five-thirds evaluations per iteration (see Exercise 2.6). On the other hand, the interval-shrinking factor is $\frac{1}{3}$ compared with $\frac{1}{2}$ for bisection. It can be shown (Exercise 2.7) that for a prescribed accuracy, we have

$$\frac{\text{total computational work in trisection}}{\text{total computational work in bisection}} = \frac{\left(\frac{5}{3}\right)\log 2}{\log 3} \cong 1.0515$$

so that bisection actually enjoys a 5 percent edge over trisection. The good students should be asked to prove that division of the interval into k equal parts exhibits a monotonically decreasing efficiency (of the root-finding procedure) with increasing k. These efficiency-balance considerations, of course, change significantly for a computer that performs numerous evaluations *in parallel*.

Returning to the bisection algorithm and remembering that the original interval $[a, b]$ is of unit length, because $1 \leq \sqrt{s} < 2$, we see that after n iterations the current approximation x_n satisfies

$$|x_n - \sqrt{s}| \leq \left(\frac{1}{2}\right)^n \tag{2.4}$$

For example, for $n = 31$ we find that $|x_{31} - \sqrt{s}| < \frac{1}{2} \cdot 10^{-9}$, so nine correct figures are assured (see Exercise 2.8). This gives us an idea of the computational work involved when the bisection method is used and provides a yardstick for comparing it with more sophisticated methods later.

One of the intrinsic properties of the bisection method is that the error bound for each iteration equals half the bound obtained for the previous iteration, which is precisely where (2.4) comes from. It is only natural, then, to look for iterations leading to $\sqrt{s}$, in which the error bound will equal, say, the square of the previous bound. If the latter bound is smaller than unity, a considerable improvement in the computation of $\sqrt{s}$ will result, as will be seen next.

2.4 Second-order iterations

Our immediate aim is the construction of iterations satisfying

$$|x_{n+1} - \sqrt{s}| \leq |x_n - \sqrt{s}|^2 \tag{2.5}$$

where the iterant x_n is the approximation to $\sqrt{s}$ after the nth iteration. However, we first assess the implications of the property (2.5), compared with the quality of approximation to $\sqrt{s}$ possessed by the bisection method, as reflected by (2.4).

Since $1 \leq s < 4$, $1 \leq \sqrt{s} < 2$, we can use the *initialization* (choice of an initial value)

$$x_0 = 1.5 \tag{2.6}$$

assuring the initial inequality

$$|x_0 - \sqrt{s}| \leq \frac{1}{2} \tag{2.7}$$

It follows now from (2.5)–(2.7) that for such a process (if indeed we can construct it), we have

$$\begin{aligned}
|x_1 - \sqrt{s}| &\leq |x_0 - \sqrt{s}|^2 \leq \left(\frac{1}{2}\right)^2 \\
|x_2 - \sqrt{s}| &\leq |x_1 - \sqrt{s}|^2 \leq \left(\frac{1}{2}\right)^{2^2} \\
|x_3 - \sqrt{s}| &\leq |x_2 - \sqrt{s}|^2 \leq \left(\frac{1}{2}\right)^{2^3}
\end{aligned} \tag{2.8}$$

and so on, and in general,

$$|x_n - \sqrt{s}| \leq \left(\frac{1}{2}\right)^{2^n} \tag{2.9}$$

Comparing this error estimate with the estimate $(\frac{1}{2})^n$ for the bisection method shows that a considerable improvement has been achieved. For example, as we saw, 31 bisection iterations were required to guarantee nine correct figures, whereas according to (2.9), five iterations suffices. This follows from the inequality $(\frac{1}{2})^{32} < \frac{1}{2}\cdot 10^{-9}$. In other words, for typical microcomputer accuracy (let alone double precision), a few dozen bisection iterations were needed, compared with a mere five or six iterations possessing the property (2.5). This property guarantees that the number of correct digits is actually doubled in each iteration, which explains why fewer iterations suffice. An iterative process with the property (2.5) is said to be *second order*, or to possess *quadratic behavior*. In this connection see Exercise 2.9. Thus, the search for second-order iterations is highly motivated in view of the expected payoff.

Our approach is the reverse of the usual practice, in which certain iterative procedures are suggested and only then analyzed for their order and other properties. That is, we propose to construct iterative algorithms that are *predesigned* to be of second order.

With a view toward this goal, we try to determine an expression p_n such that the equality

$$x_{n+1} - \sqrt{s} = \frac{(x_n - \sqrt{s})^2}{p_n} \tag{2.10}$$

will lead to the desired second-order iterative formula, in which x_{n+1} is expressed in terms of x_n and s. We rewrite (2.10) in the form

$$x_{n+1} = \frac{x_n^2 + s}{p_n} + \left(1 - \frac{2x_n}{p_n}\right)\sqrt{s} \tag{2.11}$$

from which we see that (only) the choice $p_n = 2x_n$ will rid us of $\sqrt{s}$. Obviously, we must not use $\sqrt{s}$ in the iterative formula because its value is unknown. Moreover, the computation of $\sqrt{s}$ is the very reason we set up the iterative process in the first place. The choice $p_n = 2x_n$ in (2.11) yields $x_{n+1} = (x_n^2 + s)/(2x_n)$, that is, the iterative formula

$$x_{n+1} = \frac{1}{2}\left(x_n + \frac{s}{x_n}\right) \tag{2.12}$$

which, according to (2.10), has the built-in property

$$x_{n+1} - \sqrt{s} = \frac{(x_n - \sqrt{s})^2}{2x_n} \tag{2.13}$$

The laboratory instructor should point out that (2.12) was obtained by Heron of Alexandria (about 100 B.C.) by a different approach. Using methods of differential calculus, Newton and Raphson arrived at the same result as a special case of their method of tangents. A method based on the comparison of the areas of a rectangle and a square can also be used to reach (2.12). All these approaches lead to some iterative formula, which is a posteriori shown to be of second order. As mentioned before, our approach goes precisely in the opposite direction in that it advocates construction of second-order iterations by a priori design. We shall adhere to this approach from now on to obtain sharper results in a natural way.

We infer from (2.13) that if x_0 is positive, then $x_1, x_2, \ldots$ are all positive and, moreover, $x_k \geq \sqrt{s} \geq 1$ for all $k \geq 1$. In other words, our approximations to $\sqrt{s}$ are approximations from above.

Since $1 \leq \sqrt{s} < 2$, we choose x_0 from the interval $[1,2]$; for example, $x_0 = 1.5$, in which case $|x_0 - \sqrt{s}| \leq \frac{1}{2}$. Thus, $x_n \geq 1$ for all n, including $n = 0$, and we have

$$\frac{1}{2x_n} \leq \frac{1}{2} \tag{2.14}$$

It follows now from (2.13) and (2.14) that

$$|x_{n+1} - \sqrt{s}| \leq \frac{1}{2x_n}|x_n - \sqrt{s}|^2 \leq \frac{1}{2}|x_n - \sqrt{s}|^2 \tag{2.15}$$

This, of course, implies (2.5) – our original point of departure – which eventually leads us to the error estimate (2.9), $|x_n - \sqrt{s}| \leq (\frac{1}{2})^{2^n}$. Moreover, the estimate (2.15), being sharper than (2.5), actually yields an error estimate that is even sharper than (2.9), as we shall see presently (see also Exercise 2.10).

The inequalities (2.5) and (2.15) lead us in a natural way to define a second-order iterative process for the computation of $\sqrt{s}$ as a process that, for some constant $K > 0$, satisfies

$$|x_{n+1} - \sqrt{s}| \leq K|x_n - \sqrt{s}|^2 \tag{2.16}$$

Let us investigate the consequences of (2.16), assuming that the initialization of the iterative process satisfies

$$|x_0 - \sqrt{s}| \leq \lambda \tag{2.17}$$

where λ is a constant about which we shall have more to say later on (we recall that $\lambda = \frac{1}{2}$ for $x_0 = 1.5$). Now we combine (2.16) and (2.17), and obtain

$$\begin{aligned} |x_1 - \sqrt{s}| &\leq K|x_0 - \sqrt{s}|^2 \leq K\lambda^2 \\ |x_2 - \sqrt{s}| &\leq K(K\lambda^2)^2 \leq K^{1+2}\lambda^{2^2} \\ |x_3 - \sqrt{s}| &\leq K(K^{1+2}\lambda^{2^2})^2 \leq K^{1+2+2^2}\lambda^{2^3} \end{aligned} \tag{2.18}$$

and so on, and in general,

$$|x_n - \sqrt{s}| \leq K^{1+2+2^2+\cdots+2^{n-1}} \lambda^{2^n} \tag{2.19}$$

Using the formula for the sum of a geometric progression, we have

$$1 + 2 + 2^2 + \cdots + 2^{n-1} = \frac{2^n - 1}{2 - 1} = 2^n - 1 \tag{2.20}$$

and, writing $K^{2^n - 1} = (1/K)K^{2^n}$, we reach the error estimate

$$|x_n - \sqrt{s}| \leq \frac{1}{K}(K\lambda)^{2^n} \tag{2.21}$$

For $K = 1$ and the initialization (2.7), for which $\lambda = \frac{1}{2}$, (2.21) reduces to (2.9), whereas the same initialization together with (2.15), for which $K = \frac{1}{2}$, yields (see also Exercise 2.10) the sharper error estimate

$$|x_n - \sqrt{s}| \leq 2\left(\frac{1}{4}\right)^{2^n} \tag{2.22}$$

Accordingly, four iterations of (2.12) suffice to ensure nine correct figures whereas the estimate (2.9) showed that five iterations would be needed to achieve the same accuracy.

A close look at the error estimate (2.21) reveals the sensitivity of the quality of approximations to the initial value x_0, reflected by the size of λ in (2.17). It is worthwhile, therefore, to look for an improved initialization x_0, for which the corresponding value of λ is smaller than the value $\lambda = \frac{1}{2}$ associated with $x_0 = 1.5$. We turn to this issue next.

2.5 Improved initialization

To achieve an improved initialization for the iterative computation of $\sqrt{s}$, the instructor should suggest to the laboratory participants the following idea. Consider the graph of $y = \sqrt{s}$ in the relevant interval [1, 4], and draw the secant line connecting the points $(1, 1)$ and $(4, 2)$ as in Fig. 2.2. The equation of this line is

$$y = \frac{1}{3}s + \frac{2}{3} \tag{2.23}$$

and clearly $s/3 + 2/3 \leq \sqrt{s}$ for $1 \leq s \leq 4$. It follows that (2.23) supplies us with an *estimate from below* of $\sqrt{s}$ in $[1, 4]$. To obtain an *estimate from above* as well, let us also draw the tangent to the graph of $y = \sqrt{s}$ parallel to the secant (2.23), as shown in Fig. 2.2. Because the tangent

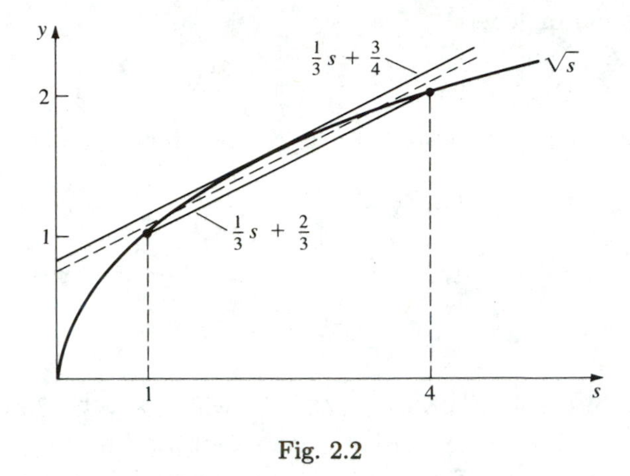

Fig. 2.2

and the secant have equal slopes, the equation of the tangent line appears as

$$y = \frac{1}{3}s + \beta \tag{2.24}$$

The value of β is determined by the requirement that the line (2.24) be tangent to the graph of $y = \sqrt{s}$, that is, that they have a unique point in common. This is equivalent to the requirement that the equation

$$\sqrt{s} = \frac{1}{3}s + \beta \tag{2.25}$$

must have only one (double) solution. If we set $u = \sqrt{s}$, we may rewrite this equation in the form

$$u^2 - 3u + 3\beta = 0 \tag{2.26}$$

and demand that its discriminant vanish. This yields $\beta = \frac{3}{4}$, and (2.24) becomes

$$y = \frac{1}{3}s + \frac{3}{4} \tag{2.27}$$

supplying us with the desired estimate from above. Using both (2.23) and (2.27), we can write, for $1 \leq s \leq 4$,

$$\frac{1}{3}s + \frac{2}{3} \leq \sqrt{s} \leq \frac{1}{3}s + \frac{3}{4} \tag{2.28}$$

It follows that we may choose either the lower initialization

$$x_0 = \frac{1}{3}s + \frac{2}{3} \tag{2.29}$$

or the upper initialization

$$x_0 = \frac{1}{3}s + \frac{3}{4} \tag{2.30}$$

The subtraction of x_0 from each of the three expressions in (2.28) ensures that for either of those two choices of x_0, we have

$$|x_0 - \sqrt{s}| \leq \frac{3}{4} - \frac{2}{3} = \frac{1}{12} \tag{2.31}$$

for $1 \leq s \leq 4$. Accordingly, we can set $\lambda = \frac{1}{12}$ in (2.17), which represents a considerable improvement over the value $\lambda = \frac{1}{2}$ associated with the simple initialization $x_0 = 1.5$. Having gone this far, we now obtain a further improvement by drawing a line parallel to the tangent and the secant that passes exactly midway between the two (the dotted line in Fig. 2.2). This line is given by

$$y = \frac{1}{3}s + \frac{17}{24} \tag{2.32}$$

because its slope equals that of the tangent and the secant, and its y-intercept averages theirs. The very location of the line (2.32) implies that the improved initialization

$$x_0 = \frac{1}{3}s + \frac{17}{24} \tag{2.33}$$

guarantees that

$$|x_0 - \sqrt{s}| \leq \frac{1}{24} \tag{2.34}$$

for $1 \leq s \leq 4$. Alternatively, we could have obtained (2.34) by subtracting (2.33) from each of the three expressions in (2.28). Using this improved initialization, for which $\lambda = \frac{1}{24}$, and recalling from the last section that $K = \frac{1}{2}$, we find that the error decay (2.21) is actually given by

$$|x_n - \sqrt{s}| \leq 2\left(\frac{1}{48}\right)^{2^n} \tag{2.35}$$

With the improved initialization (2.33), we find that a mere three iterations of (2.12) will yield 12 correct figures.

It is worthwhile explaining the motivation behind the effort we are making to reduce the number of iterations, and thus the computational work, by even a small amount. We must not lose sight of the fact that our aim is to lay the foundations of a *computer library function* for the computation of square roots. This library function will be invoked a vast number of times, so any computational saving, however small, contributes significantly to the overall efficiency.

At this point the instructor should suggest to the better laboratory participants to search for an even more improved initialization, topping (2.33). In this connection see Exercises 2.11 and 2.12.

Numerical results obtained with the bisection and second-order methods (using simple as well as improved initializations) for the computation of $\sqrt{3}$, are given in Table 3.1 in the next section.

2.6 Third-order iterations

In the spirit of our construction of a second-order iterative method for the computation of $\sqrt{s}$, we are naturally led to try a third-order algorithm, which will triple the number of correct digits in each iteration. In analogy to (2.10), we attempt to determine an expression q_n such that the equality

$$x_{n+1} - \sqrt{s} = \frac{(x_n - \sqrt{s})^3}{q_n} \tag{2.36}$$

leads to the desired third-order iterative formula. We rewrite this equality in the form

$$x_{n+1} = \frac{x_n^3 + 3x_n s}{q_n} + \left(1 - \frac{3x_n^2 + s}{q_n}\right)\sqrt{s} \tag{2.37}$$

from which we see that (only) the choice $q_n = 3x_n^2 + s$ will rid us of $\sqrt{s}$, as is obviously required. This choice yields the third-order iterative formula

$$x_{n+1} = x_n \frac{x_n^2 + 3s}{3x_n^2 + s} \tag{2.38}$$

which, according to (2.36), has the built-in property

$$x_{n+1} - \sqrt{s} = \frac{(x_n - \sqrt{s})^3}{3x_n^2 + s} \tag{2.39}$$

It follows that if $x_0 \geq \sqrt{s}$, then $x_k \geq \sqrt{s}$ for all k (approximations from above), whereas if $x_0 \leq \sqrt{s}$, then $x_k \leq \sqrt{s}$ for all k (approximations from below). This behavior is different from that of the second-order, where the approximations are always found to be from above. For the third-order iterations, though, we find that in all cases we have

$$|x_{n+1} - \sqrt{s}| = \frac{|x_n - \sqrt{s}|^3}{3x_n^2 + s} \leq \frac{1}{s}|x_n - \sqrt{s}|^3 \tag{2.40}$$

Since $\sqrt{s} \geq 1$, we obtain the typical third-order behavior

$$|x_{n+1} - \sqrt{s}| \leq |x_n - \sqrt{s}|^3 \tag{2.41}$$

Now, for the initialization x_0 satisfying $|x_0 - \sqrt{s}| \leq \lambda$, a repeated substitution of the inequality (2.41) into itself (see Exercise 2.13) leads to

$$|x_n - \sqrt{s}| \leq \lambda^{3^n} \tag{2.42}$$

If, in conjunction with (2.42), we use the improved initialization $x_0 = (8s + 17)/24$, with the associated $\lambda = \frac{1}{24}$, then we have the estimate

$$|x_n - \sqrt{s}| \leq \left(\frac{1}{24}\right)^{3^n} \tag{2.43}$$

Thus two iterations of (2.38) guarantee 12 correct figures.

At first sight, the third-order iterations appear to be much more efficient than those of the second-order ones, because fewer iterations are needed for a prescribed accuracy. We must remember, however, that this gain is almost wiped out by the additional computational complexity per third-order iteration. On balance, the advantage of the third-order method is rather marginal, and for this reason there seems to be no practical advantage in the construction of fourth- and higher-order methods (see, however, Exercise 2.14).

To demonstrate the performance of the various methods discussed so far, we ran a program that computes $\sqrt{3}$, using bisection as well as second- and third-order methods, along with the simple initialization $x_0 = 1.5$ and the improved initialization $x_0 = (8s + 17)/24$, for $s = 3$. Table 2.1 compares the quality of approximation of the various methods by displaying the deviations of our results from a very accurate value of $\sqrt{3}$. The table demonstrates the enormous advantage of second- and third-order methods over bisection, as well as the superiority of the improved initialization. It can also be seen that the third-order method enjoys only a marginal advantage over the second-order method, hardly compensating for the increased computational price incurred per third-order iteration. At this point, the laboratory participants must write their own programs for the computation of $\sqrt{s}$, using all the methods described (see Exercise 2.15). This is the time and place to consummate the numerical laboratory approach, in which the laboratory participants (aided by the instructor) can see for themselves that mathematics really works.

2.7 Cube roots

We now turn our attention to the construction of algorithms for efficient computations of cube roots. Let us recall the binary representation of a

Table 2.1. *Deviations from* $\sqrt{3} = 1.73205080757$

n	Bisection	Second order (simple)	Second order (improved)	Third order (simple)	Third order (improved)
1	−.232050808	.017949192	.000164640	−.001282526	−.000001123
2	+.017949192	.000092050	.000000008	−.000000001	−.000000000
3	−.107050808	.000000002	.000000000	−.000000000	
4	−.044550808	.000000000			
5	−.013300808				
6	+.002324192				
7	−.005488308				
8	−.001582058				
9	+.000371067				
10	−.000605495				
11	−.000117214				
12	+.000126927				
13	+.000004856				
14	−.000056179				
15	−.000025661				
16	−.000010402				
17	−.000002773				
18	+.000001042				
19	−.000000866				
20	+.000000088				
21	−.000000389				
22	−.000000150				
23	−.000000031				
24	+.000000029				
25	−.000000002				
26	+.000000014				
27	+.000000006				
28	+.000000002				
29	+.000000001				
30	+.000000000				

positive number d in the form $d = m \cdot 2^i$, which enables us (see Exercise 2.1) to cast d into the equivalent form
$d = s \cdot 8^j$, where

$$\begin{cases} s = m, \quad j = \dfrac{i}{3}, & \text{when } \; i \text{ is divisible by 3} \\ s = 2m, \quad j = \dfrac{(i-1)}{3}, & \text{when } \; (i-1) \text{ is divisible by 3} \\ s = 4m, \quad j = \dfrac{(i-2)}{3}, & \text{when } \; (i-2) \text{ is divisible by 3} \end{cases} \tag{2.44}$$

Here $1 \leq s < 8$ since $1 \leq m < 2$, and taking cube roots we reach

$$\sqrt[3]{d} = \sqrt[3]{s} \cdot 2^j \tag{2.45}$$

We have therefore again reduced the range of numbers, whose cube roots we may want to compute, to $[1, 8)$. Moreover, the result (here, $\sqrt[3]{s}$) is again in $[1, 2)$.

Next, let us attempt to construct second-order iterations for cube roots in a way analogous to that for square roots. Accordingly, we try to determine p_n so that the equality

$$x_{n+1} - \sqrt[3]{s} = \frac{(x_n - \sqrt[3]{s})^2}{p_n} \tag{2.46}$$

will yield the desired iterative formula. This equality can be written in the form

$$x_{n+1} = \frac{x_n^2}{p_n} + \frac{p_n - 2x_n}{p_n} \sqrt[3]{s} + \frac{1}{p_n} (\sqrt[3]{s})^2 \tag{2.47}$$

and it is clear that we must eliminate the explicit dependence on $\sqrt[3]{s}$ and $(\sqrt[3]{s})^2$. The laboratory participants will readily find that this cannot be accomplished, no matter how p_n is chosen. The reason, of course, is that we are attempting to eliminate two quantities when only one expression is at our disposal. We therefore conclude that (2.46) is too naive and suggest its generalization to

$$x_{n+1} - \sqrt[3]{s} = \frac{(x_n - \sqrt[3]{s})^2}{p_n} + \frac{(x_n - \sqrt[3]{s})^3}{q_n} \tag{2.48}$$

in which we have introduced two expressions to be determined in such a way as to attain our goal. If indeed we succeed in doing so, then we can write (2.48) in the factored form

$$x_{n+1} - \sqrt[3]{s} = \left[\frac{1}{p_n} + \frac{x_n - \sqrt[3]{s}}{q_n}\right] (x_n - \sqrt[3]{s})^2 \tag{2.49}$$

which turns out to be indispensable for the investigation of the second-order behavior of the resulting iterations.

To determine p_n and q_n, let us rewrite (2.48) in yet another form,

$$\begin{aligned} x_{n+1} = & \left(\frac{x_n^2}{p_n} + \frac{x_n^3 - s}{q_n}\right) + \left(1 - \frac{2x_n}{p_n} - \frac{3x_n^2}{q_n}\right) \sqrt[3]{s} \\ & + \left(\frac{1}{p_n} + \frac{3x_n}{q_n}\right) (\sqrt[3]{s})^2 \end{aligned} \tag{2.50}$$

Now we choose p_n and q_n so that the quantities multiplying $\sqrt[3]{s}$ and $(\sqrt[3]{s})^2$ vanish, that is,

$$\begin{aligned} \frac{2x_n}{p_n} + \frac{3x_n^2}{q_n} &= 1 \\ \frac{1}{p_n} + \frac{3x_n}{q_n} &= 0 \end{aligned} \tag{2.51}$$

The unique solution of (2.51) is $p_n = x_n$ and $q_n = -3x_n^2$, which reduces (2.50) to $x_{n+1} = (2x_n^3 + s)/(3x_n^2)$ or

$$x_{n+1} = \frac{2}{3}\left(x_n + \frac{s/2}{x_n^2}\right) \tag{2.52}$$

The laboratory participants should note that the numbers $2/3$ and $s/2$ are to be computed before iterating, so that each iteration consumes one addition, two multiplications, and one division.

Now we substitute $p_n = x_n$ and $q_n = -3x_n^2$ in the factored formula (2.49) to obtain

$$x_{n+1} - \sqrt[3]{s} = \left(\frac{2}{3x_n} + \frac{\sqrt[3]{s}}{3x_n^2}\right)(x_n - \sqrt[3]{s})^2 \tag{2.53}$$

We shall always choose x_0 to be positive (in fact, $x_0 \geq 1$), so it follows from (2.53) that $x_1 \geq \sqrt[3]{s} \geq 1$ because the right-hand side is nonnegative. Repeating the argument, we find that $x_2 \geq \sqrt[3]{s} \geq 1$, and so on. Hence $x_n \geq 1$ for all n, including $n = 0$. Accordingly,

$$|x_{n+1} - \sqrt[3]{s}| \leq \left(\frac{2}{3} + \frac{\sqrt[3]{s}}{3}\right)|x_n - \sqrt[3]{s}|^2 \leq \frac{4}{3}|x_n - \sqrt[3]{s}|^2 \tag{2.54}$$

because $\sqrt[3]{s} < 2$. This completes our verification of the built-in second-order behavior of the iterations (2.52).

We turn now to the issue of initializing the cube-root iterations. Thus, suppose that the initial value x_0 satisfies

$$|x_0 - \sqrt[3]{s}| \leq \lambda \tag{2.55}$$

For example, the simple initialization $x_0 = 1.5$ guarantees $\lambda = \frac{1}{2}$, whereas an improved initialization, to be discussed, will lead to a smaller value of λ. In all cases, the estimate (2.54) yields

$$|x_1 - \sqrt[3]{s}| \leq \frac{4}{3}\lambda^2 \tag{2.56}$$

which can be inserted into (2.54) to get a bound on $|x_2 - \sqrt[3]{s}|$, and so on. Continuing this process, we eventually reach

$$|x_n - \sqrt[3]{s}| \leq \frac{3}{4}\left(\frac{4}{3}\lambda\right)^{2^n} \tag{2.57}$$

Thus, even $\lambda = \frac{1}{2}$ provides us with an error bound that decays at a rate proportional to $(\frac{2}{3})^{2^n}$ and guarantees 11 correct figures after only 6 iterations of (2.52). The better students are challenged, at this point, to work out Exercises 2.16 and 2.17.

As has been the case for square roots, we see from (2.57) that the cube-root approximations are highly sensitive to the quality of the initialization. It is natural, therefore, to construct again an improved initialization, reducing the value of λ. It is suggested that the instructor guide the laboratory participants along lines analogous to those employed in Section 2.5 for $\sqrt{s}$. This time we consider the graph of $y = \sqrt[3]{s}$ in the relevant interval $[1, 8]$ and again draw the secant line connecting (in this case) the points $(1, 1)$ and $(8, 2)$, as in Fig. 2.3. The equation of this line is found to be

$$y = \frac{1}{7}s + \frac{6}{7} \tag{2.58}$$

and clearly $s/7 + 6/7 \leq \sqrt[3]{s}$ for $1 \leq s \leq 8$. It follows that (2.58) supplies us with an estimate from below of $\sqrt[3]{s}$ in $[1, 8]$. To obtain an estimate

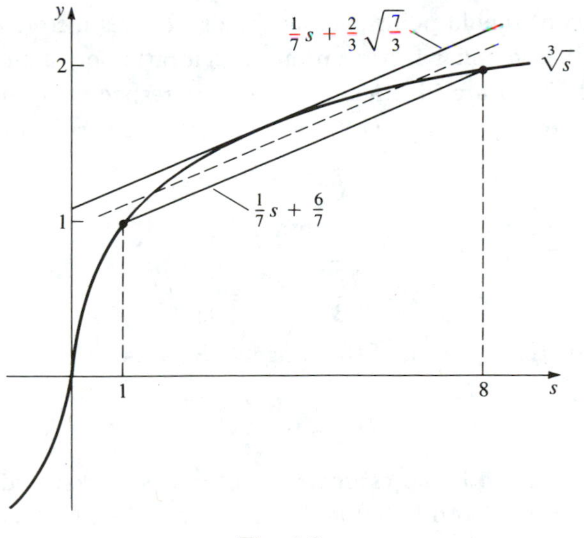

Fig. 2.3

from above, we also draw the tangent to the graph of $y = \sqrt[3]{s}$, parallel to the secant. The equation of this tangent line is clearly of the form

$$y = \frac{1}{7}s + \beta \tag{2.59}$$

with β to be determined by the requirement that the equation

$$\frac{1}{7}s + \beta = \sqrt[3]{s} \tag{2.60}$$

have two identical positive solutions (see Fig. 2.3). In other words, the line (2.59) is the desired tangent if, for $s > 0$, it has a unique point in common with the graph of $y = \sqrt[3]{s}$. This geometrical condition determines the value of β.

For simplicity, we set $u = \sqrt[3]{s}$ and rewrite equation (2.60) in the form

$$u^3 - 7u + 7\beta = 0 \tag{2.61}$$

The geometrical condition of tangency is now equivalent to the requirement that (2.61) have a double positive root, say at $u = \rho$ (the third root, $u = \mu$, is negative, as will be seen). This requirement means that (2.61) can be factored in the form

$$(u - \rho)^2(u - \mu) = 0 \tag{2.62}$$

that is,

$$u^3 - (2\rho + \mu)u^2 + \rho(\rho + 2\mu)u - \rho^2\mu = 0 \tag{2.63}$$

The instructor should point out the analogy of this factorization to the corresponding, familiar factorization of quadratic polynomials. Because (2.61) and (2.63) are identical, so are their respective coefficients. An immediate consequence is that $\mu = -2\rho$, and (2.63) reduces to

$$u^3 - 3\rho^2 u + 2\rho^3 = 0 \tag{2.64}$$

Comparing this with (2.61), we obtain

$$\rho = \sqrt{\frac{7}{3}}, \qquad \beta = \frac{2}{3}\sqrt{\frac{7}{3}} \tag{2.65}$$

Accordingly, the equation of the tangent (2.59) is

$$y = \frac{1}{7}s + \frac{2}{3}\sqrt{\frac{7}{3}} \tag{2.66}$$

and supplies us with the estimate from above to $\sqrt[3]{s}$, which we were looking for. Using both (2.58) and (2.66), we can write, for $1 \le s \le 8$,

$$\frac{1}{7}s + \frac{6}{7} \le \sqrt[3]{s} \le \frac{1}{7}s + \frac{2}{3}\sqrt{\frac{7}{3}} \tag{2.67}$$

In analogy to what we did in Section 2.5, we draw the line parallel to the tangent and the secant that passes exactly midway between the two (the dotted line in Fig. 2.3). This line is given by

$$y = \frac{1}{7}s + \left(\frac{3}{7} + \frac{1}{3}\sqrt{\frac{7}{3}}\right) \tag{2.68}$$

as can easily be verified, and its very location assures us that the improved initialization

$$x_0 = \frac{1}{7}s + \left(\frac{3}{7} + \frac{1}{3}\sqrt{\frac{7}{3}}\right) \tag{2.69}$$

guarantees that

$$|x_0 - \sqrt[3]{s}| \le \frac{1}{2}\left(\frac{2}{3}\sqrt{\frac{7}{3}} - \frac{6}{7}\right) < \frac{1}{12} \tag{2.70}$$

for $1 \le s \le 8$. This estimate can also be obtained by subtracting (2.69) from each of the three expressions in (2.67). The numerical coefficients in the initialization (2.69) are to be computed ahead of time and stored in the computer library function for cube roots. Using this improved initialization, for which we can take $\lambda = \frac{1}{12}$ in the error estimate (2.57), we obtain

$$|x_n - \sqrt[3]{s}| < \frac{3}{4}\left(\frac{1}{9}\right)^{2^n} \tag{2.71}$$

It follows, for example, that with the improved initialization, four iterations of (2.52) yield 15 correct figures.

Just as we constructed third-order iterations for the computation of $\sqrt{s}$, we can develop a third-order method for $\sqrt[3]{s}$. We indicate only the essentials. In analogy to (2.48), we start with the equality

$$x_{n+1} - \sqrt[3]{s} = \frac{(x_n - \sqrt[3]{s})^3}{p_n} + \frac{(x_n - \sqrt[3]{s})^4}{q_n} \tag{2.72}$$

and eventually reach the iterative formula

$$x_{n+1} = \frac{1}{2}\left(x_n + \frac{3sx_n}{2x_n^3 + s}\right) \tag{2.73}$$

Moreover, it can be shown that these iterations satisfy $|x_{n+1} - \sqrt[3]{s}| \le |x_n - \sqrt[3]{s}|^3$, and hence $|x_n - \sqrt[3]{s}| \le \lambda^{3^n}$. In this connection see Exercises 2.18 and 2.19.

Typical results for the computation of cube roots, using second- and third-order methods, with both simple and improved initializations, are

Table 2.2. *Deviations from* $\sqrt[3]{5} = 1.70997594668$

n	Second order (simple)	Second order (improved)	Third order (simple)	Third order (improved)
1	.030764795	.002056092	−.002529137	−.000046659
2	.000540516	.000002469	−.000000003	−.000000000
3	.000000171	.000000000	−.000000000	
4	.000000000			

shown in Table 2.2 for $s = 5$. To demonstrate the quality of the various approximations, we display the deviations of our results from a very accurate value of $\sqrt[3]{5}$. Table 2.2 shows again the superiority of the improved initialization and the fact that the third-order method is only marginally better than the second-order method. The negative deviations of the third-order approximations in Table 2.2 (and Table 2.1) are because the initialization is from below ($x_0 < \sqrt[3]{5}$), causing all subsequent approximations to be from below as well ($x_n \leq \sqrt[3]{5}$). Had we chosen to employ an initialization from above, we would have obtained approximations x_n from above and hence positive deviations. In this connection see Exercises 2.20, 2.21, and 2.22, which also address the issue of second-order methods in which the deviations are always positive.

In a manner completely analogous to that just described, it is possible to construct iterations to compute $\sqrt[k]{s}$ for all integral k. The point of departure is an equality analogous to (2.48) or (2.72), containing $(k-1)$ terms, each of whose denominators is to be determined subsequently. In this connection see Exercise 2.23.

The iterations developed so far are not the only ones possible. Next we consider iterations of a different nature and compare them with methods already developed.

2.8 Polynomial iterations

When we constructed the second-order iterative formula for $\sqrt{s}$ in the previous sections, we started with the equality

$$x_{n+1} - \sqrt{s} = \frac{1}{p_n}(x_n - \sqrt{s})^2 \tag{2.74}$$

then obtained $p_n = 2x_n$, and subsequently the iterative formula (2.12). This formula necessitates addition, multiplication, and division. Because division consumes at least twice as much computing time as multiplication, it seems worthwhile to investigate the possibility of constructing second-order iterations containing only addition and multiplication, which are, therefore, *division-free*. To this end, we try to determine two parameters α and β so that the equality

$$x_{n+1} - \sqrt{s} = (\alpha x_n + \beta)(x_n - \sqrt{s})^2 \tag{2.75}$$

leads to the desired division-free formula. We write (2.75) in the form

$$x_{n+1} = \alpha x_n^3 + (\beta - 2\alpha\sqrt{s})x_n^2 + (\alpha s - 2\beta\sqrt{s})x_n + s\left(\beta + \frac{1}{\sqrt{s}}\right) \tag{2.76}$$

To eliminate the dependence on $\sqrt{s}$ in the last term, we must choose $\beta = -1/\sqrt{s}$, which in turn leaves us with

$$x_{n+1} = \alpha x_n^3 - \frac{1 + 2\alpha s}{\sqrt{s}} x_n^2 + (2 + \alpha s)x_n \tag{2.77}$$

It follows that (only) the choice $\alpha = -1/(2s)$ will rid us of $\sqrt{s}$ and yield the iterative formula

$$x_{n+1} = x_n \left(\frac{3}{2} - \frac{1}{2s}x_n^2\right) \tag{2.78}$$

The numbers $3/2$ and $-1/(2s)$ should be precomputed once, before iterating, so that the iterations (2.78) are indeed division-free. This is what we set out to find, and because the right-hand side of (2.78) is a polynomial in x_n, it will be referred to as a *polynomial iterative formula*. The laboratory participants should verify for themselves that the use of only one parameter (either α or β) would not have sufficed and that the only two-parameter choice is the one just obtained.

Next, to examine the behavior of our polynomial iterative formula, we substitute $\alpha = -1/(2s)$, $\beta = -1/\sqrt{s}$ into (2.75). This yields

$$x_{n+1} - \sqrt{s} = -\left(\frac{x_n}{2s} + \frac{1}{\sqrt{s}}\right)(x_n - \sqrt{s})^2 \tag{2.79}$$

In addition, we rewrite (2.78) in the form

$$x_{n+1} = x_n + \frac{x_n}{2s}(s - x_n^2) \tag{2.80}$$

Now, if $0 < x_n \leq \sqrt{s}$, then (2.80) shows that $x_n \leq x_{n+1}$ whereas (2.79) assures us that $x_{n+1} \leq \sqrt{s}$. Hence, if we start iterations with an initialization from below, satisfying $0 < x_0 \leq \sqrt{s}$, we will have an increasing bounded sequence x_n, satisfying

$$0 < x_n \leq \sqrt{s} \tag{2.81}$$

for all n. Using this result in (2.79), we obtain

$$\frac{|x_{n+1} - \sqrt{s}|}{|x_n - \sqrt{s}|^2} = \frac{1}{\sqrt{s}}\left[\frac{1}{2}\frac{x_n}{\sqrt{s}} + 1\right] \leq \frac{\frac{3}{2}}{\sqrt{s}} \tag{2.82}$$

and since $1 \leq \sqrt{s} < 2$, we have the second-order behavior

$$|x_{n+1} - \sqrt{s}| \leq \frac{3}{2}|x_n - \sqrt{s}|^2 \tag{2.83}$$

Having worked with initialization from below, we invoke (2.29) and choose $x_0 = (s+2)/3$, which ensures that

$$|x_0 - \sqrt{s}| \leq \frac{1}{12} = \lambda \tag{2.84}$$

Recalling (2.16) and (2.21), we find that at present $K = \frac{3}{2}$, and hence

$$|x_n - \sqrt{s}| \leq \frac{1}{K}(K\lambda)^{2^n} = \frac{2}{3}\left(\frac{1}{8}\right)^{2^n} \tag{2.85}$$

Accordingly, a mere four polynomial iterations guarantees 14 correct figures. We remark that had we used a simple lower initialization, such as

$$x_0 = \begin{cases} 1.0, & 1.00 \leq s < 2.25 \\ 1.5, & 2.25 \leq s < 4.00 \end{cases} \tag{2.86}$$

for which $\lambda = \frac{1}{2}$, we would have obtained 15 correct figures after seven iterations (see Exercise 2.24).

With a view toward constructing polynomial iterations of third order for $\sqrt{s}$, and in the spirit of (2.75), we start with the three-parameter equality

$$x_{n+1} - \sqrt{s} = (\alpha x_n^2 + \beta x_n + \gamma)(x_n - \sqrt{s})^3 \tag{2.87}$$

and try to determine α, β, and γ judiciously. It can be verified that two parameters are insufficient for the purpose and, at the same time, that the unique values of α, β, and γ, leading to third-order polynomial iterations are (see Exercise 2.25)

$$\alpha = \frac{3}{8s^2}, \qquad \beta = \frac{9}{8s\sqrt{s}}, \qquad \gamma = \frac{1}{s} \tag{2.88}$$

Using these values in (2.87), we reach the desired polynomial iterative formula

$$x_{n+1} = x_n\left[\left(\frac{3}{8s^2}x_n^2 - \frac{10}{8s}\right)x_n^2 + \frac{15}{8}\right] \tag{2.89}$$

in which the numbers $3/(8s^2)$, $-10/(8s)$, and $15/8$ should be precomputed once before iterating. This formula can also be cast into the form

$$x_{n+1} = x_n + \frac{3}{8s^2}(s - x_n^2)\left(\frac{7}{3}s - x_n^2\right) \tag{2.90}$$

from which we see that if $0 < x_n \leq \sqrt{s}$, then $x_n \leq x_{n+1}$. On the other hand, introducing the values of α, β, and γ just obtained into the original equality (2.87), we find

$$x_{n+1} - \sqrt{s} = \left(\frac{3}{8s^2}x_n^2 + \frac{9}{8s\sqrt{s}}x_n + \frac{1}{s}\right)(x_n - \sqrt{s})^3 \tag{2.91}$$

Now, if $0 < x_n \leq \sqrt{s}$, then (2.91) implies that $x_{n+1} \leq \sqrt{s}$. Because also $x_n \leq x_{n+1}$, as we have just seen, it follows that a lower initialization ($0 < x_0 \leq \sqrt{s}$) leads again to an increasing bounded sequence x_n, satisfying

$$0 < x_n \leq \sqrt{s} \tag{2.92}$$

for all n. Using this result in (2.91), we obtain

$$\frac{|x_{n+1} - \sqrt{s}|}{|x_n - \sqrt{s}|^3} \leq \frac{3}{8s} + \frac{9}{8s} + \frac{1}{s} = \frac{5}{2s} \tag{2.93}$$

and since $1 \leq \sqrt{s} < 2$, we have the third-order behavior

$$|x_{n+1} - \sqrt{s}| \leq \frac{5}{2}|x_n - \sqrt{s}|^3 \tag{2.94}$$

This is a special case of the third-order behavior $|x_{n+1} - \sqrt{s}| \leq K|x_n - \sqrt{s}|^3$, with $K = \frac{5}{2}$. Let us again choose the lower initialization $x_0 = (s+2)/3$, for which $|x_0 - \sqrt{s}| \leq \frac{1}{12} = \lambda$. In general, we have for third-order iterations,

$$\begin{aligned}
&|x_0 - \sqrt{s}| \leq \lambda \\
&|x_1 - \sqrt{s}| \leq K\lambda^3 \\
&|x_2 - \sqrt{s}| \leq K(K\lambda^3)^3 = K^{1+3}\lambda^{3^2} \\
&|x_3 - \sqrt{s}| \leq K(K^{1+3}\lambda^{3^2})^3 = K^{1+3+3^2}\lambda^{3^3}
\end{aligned} \tag{2.95}$$

and so on, and generally,

$$|x_n - \sqrt{s}| \leq K^{1+3+3^2+\cdots+3^{n-1}}\lambda^{3^n} \tag{2.96}$$

Using the formula for the sum of a geometric progression, we have

$$1 + 3 + 3^2 + \cdots + 3^{n-1} = \frac{3^n - 1}{3 - 1} = \frac{1}{2}(3^n - 1) \tag{2.97}$$

Table 2.3. *Deviations from* $\sqrt{1.69} = 1.3$

n	Second order	Third order
1	−.005552367	−.000487126
2	−.000035521	−.000000000
3	−.000000002	
4	−.000000000	

and, writing $K^{(3^n-1)/2} = (1/\sqrt{K})(\sqrt{K})^{3^n}$, we reach the error estimate

$$|x_n - \sqrt{s}| \leq \frac{1}{\sqrt{K}}(\sqrt{K}\lambda)^{3^n} \tag{2.98}$$

Accordingly, with $K = \frac{5}{2}$ and $\lambda = \frac{1}{12}$, we have

$$|x_n - \sqrt{s}| \leq \sqrt{\frac{2}{5}}\left(\sqrt{\frac{5}{2}}\frac{1}{12}\right)^{3^n} < \frac{2}{3}\left(\frac{1}{7}\right)^{3^n} \tag{2.99}$$

Thus, two third-order polynomial iterations guarantee 7 correct figures, whereas 22 correct figures are assured when a third iteration is performed.

It is possible to construct second- and third-order polynomial iterations for $\sqrt[k]{s}$, $k \geq 3$ along the lines of the preceding analysis (see Exercises 2.26 and 2.27). On the other hand, as mentioned previously, there is no practical advantage in the construction of fourth- and higher-order methods.

To demonstrate the performance of the polynomial iterations just constructed, we ran a program that computes $\sqrt{1.69}$, using second- and third-order iterations, starting with the lower initialization $x_0 = (s+2)/3 = 1.23$. Note in Table 2.3 that the number of correct figures in the approximation x_n – reflected by the number of zeros in the deviations – is doubled for second-order and tripled for third-order iterations, as expected. At this point, laboratory participants are urged to write their own programs for computing $\sqrt{s}$ and $\sqrt[3]{s}$, for various values of s, using the polynomial iterative methods developed in this section (Exercises 2.28 and 2.29).

Next we look at root-finding iteration from an advanced point of view, drawing upon the knowledge of calculus. Precalculus students may skip this section without loss of continuity.

2.9 An advanced point of view

In the results obtained thus far, no use has been made of concepts from calculus. However, if laboratory participants are familiar with the basics of calculus, the instructor should point out the connection between our various iterative formulas and the (second-order) Newton–Raphson method, as well as its third-order extensions.

Suppose we want to construct a sequence of successive approximations to a solution of $f(x) = 0$. We assume that $x = \rho$ is a *simple* root of $f(x)$, that is, $f(\rho) = 0,\ f'(\rho) \neq 0$. Then the Newton–Raphson iterative formula is

$$x_{n+1} = x_n - \frac{f_n}{f'_n} \tag{2.100}$$

where $f_n = f(x_n)$ and $f'_n = f'(x_n)$. Under suitable conditions, which are omitted here, it can be shown that the approximations x_n converge to the desired root ρ, the convergence being of second order. The two most widely used third-order methods, which are offshoots of the Newton–Raphson method, are given, respectively, by Halley's method

$$x_{n+1} = x_n - \frac{2f_n f'_n}{2(f'_n)^2 - f_n f''_n} \tag{2.101}$$

and the extended Newton method

$$x_{n+1} = x_n - \frac{f_n}{f'_n} - \frac{f_n^2 f''_n}{2(f'_n)^3} \tag{2.102}$$

If we apply (2.100) to the function $f(x) = x^2 - s$, we obtain

$$x_{n+1} = \frac{1}{2}\left(x_n + \frac{s}{x_n}\right) \tag{2.103}$$

which is identical to the formula (2.12), previously obtained by elementary means. Moreover, application of (2.101) to the same $f(x)$ yields

$$x_{n+1} = x_n \frac{x_n^2 + 3s}{3x_n^2 + s} \tag{2.104}$$

which is identical to (2.38). However, the use of the same $f(x)$ in conjuction with (2.102) generates an iterative formula of a kind that is not discussed in this chapter (see, however, Exercise 2.30).

The polynomial iterations developed in Section 2.8 are also derivable from the Newton–Raphson method and its offshoots. This time we apply (2.100) to the function $f(x) = 1 - s/x^2$ and obtain precisely the polynomial iterations (2.78). Moreover, application of (2.102) to the same $f(x)$ yields (see Exercise 2.31) the third-order polynomial iterations (2.89).

However, if we apply (2.101) to this $f(x)$, we do not generate a polynomial iterative formula but, surprisingly, end up (again) with (2.104), that is, (2.38). If desirable, we may replace x^2 by x^k in both functions $f(x)$ just defined, to obtain various iterative formulas for the computation of $\sqrt[k]{s}$. In this connection see Exercises 2.32 and 2.33.

In this chapter we have presented laboratory assignments whose central theme is the construction and testing of iterative algorithms for the computation of roots. These algorithms not only are of built-in second or third order, by a priori design, but have been developed through elementary algebraic means. Thus, they are perfectly suitable as a first instructive subject for a precalculus numerical laboratory.

Exercises

2.1 Show that any positive number d whose binary representation is $d = m \cdot 2^i$ ($1 \leq m < 2$ and i is an appropriate integer) may also be represented, uniquely, in the form $d = s \cdot 8^j$. Find s and j as functions of m and i, as well as the range of s. Also show that $1 \leq \sqrt[3]{s} < 2$ (see Exercise 1.7).

2.2 Repeat Exercise 2.1, using 2^k instead of 8 ($= 2^3$). Draw the appropriate conclusions relating to the computation of $\sqrt[k]{s}$. Here k denotes any positive integer.

2.3 Carry out, by hand computations, the first five iterations in the bisection method for the computation of $\sqrt{3}$. Observe that the deviation $|\sqrt{3} - \text{approximation}|$ does not necessarily decrease at every iteration. To see that this deviation eventually will become as small as you please, carry out additional iterations and study Table 2.1.

2.4 Consider the equation $f(x) = 2^x + x - 5 = 0$ and show (graphically or otherwise) that it has a unique solution, located in the interval $[1, 2]$. Write down and run a bisection algorithm that computes this root to any desired accuracy. (Hint: $f(x) = 0$ is equivalent to $2^x = 5 - x$.)

2.5 Repeat Exercise 2.4 with $f(x) = x^7 + 2x - 200 = 0$.

2.6 Construct an algorithm for the computation of $\sqrt{s}$, $1 \leq \sqrt{s} < 2$, to any desired accuracy, by trisecting the relevant interval at each iteration. Observe that for some iterations it is sufficient to compute $y = x^2 - s$ just once whereas for the others two such computations are required. Hence, conclude that the average number of evaluations of $y = x^2 - s$ per iteration is $\frac{5}{3}$, compared with one per bisection iteration.

2.7 Suppose that n bisection iterations or m trisection iterations are needed for the solution of $f(x) = 0$ to a prescribed accuracy. Show that

$$\frac{\text{total computational work in trisection}}{\text{total computational work in bisection}} \cong 1.0515$$

and conclude that bisection enjoys a 5 percent edge over trisection.

2.8 Use the inequality $1/2^{10} = 1/1024 < 10^{-3}$ to show that if $n \geq \frac{10}{3} \cdot q + 1$, then $(\frac{1}{2})^n < \frac{1}{2} \cdot 10^{-q}$.

2.9 Prove that for every pair of positive integers m and n that satisfy $m \geq \frac{10}{3} \log_{10} n$, we have $(\frac{1}{2})^{2^m} < (\frac{1}{2})^n$. This inequality demonstrates the tremendous superiority of second-order methods over bisection.

2.10 Starting from the inequality $|x_{n+1} - \sqrt{s}| \leq \frac{1}{2}|x_n - \sqrt{s}|^2$ show, in conjunction with (2.7) and (2.8), that $|x_n - \sqrt{s}| \leq 2(\frac{1}{2})^{2^{n+1}}$. Observe that this error estimate is sharper than the estimate (2.9).

2.11 Derive the equations of the secant lines connecting the pairs of points $(1, 1)$, $(2.25, 1.50)$ and $(2.25, 1.50)$, $(4, 2)$ located on the graph of $y = \sqrt{s}$. Next, derive the equations of the tangent lines that are parallel to those secants. Finally, find the equations of the middle lines, which are parallel to and midway between the secants and their associated tangents. Draw an appropriate figure, analogous to Fig. 2.2.

2.12 Suppose, in analogy to (2.33), that we choose the value of x_0 by using the two middle lines described in Exercise 2.11, the first for $1 \leq s < 2.25$ and the second for $2.25 \leq s < 4$. Show that, for all s in $[1, 4]$, we have $|x_0 - \sqrt{s}| \leq \frac{1}{80}$ and, moreover, that the error estimate is given by $|x_n - \sqrt{s}| \leq 2(\frac{1}{160})^{2^n}$.

2.13 In Section 2.6 we proved that the iterations (2.38) are of third order and satisfy $|x_{n+1} - \sqrt{s}| \leq |x_n - \sqrt{s}|^3$. If the initialization x_0 satisfies $|x_0 - \sqrt{s}| \leq \lambda$, show that $|x_n - \sqrt{s}| \leq \lambda^{3^n}$.

2.14 In analogy to (2.36), construct the fourth-order iterations

$$x_{n+1} = \frac{x_n^4 + 6sx_n^2 + s^2}{4x_n^3 + 4sx_n}$$

for the computation of $\sqrt{s}$. Prove that

$$|x_{n+1} - \sqrt{s}| \le \left(\frac{1}{8}\right) |x_n - \sqrt{s}|^4$$

and that $|x_0 - \sqrt{s}| \le \lambda$ implies $|x_n - \sqrt{s}| \le 2(\lambda/2)^{4^n}$.

2.15 Write a computer program for the computation of $\sqrt{s}$, using the second- and third-order methods described in Sections 2.4, 2.5, and 2.6, with both simple and improved initializations. Run this program using the values $s = 1.44$, 2.25, 3.00, and 3.24. Repeat the whole procedure with the fourth-order method of Exercise 2.14.

2.16 Assuming an upper initialization satisfying $0 \le x_0 - \sqrt[3]{s} \le \delta$, use (2.53) to prove that $|x_{n+1} - \sqrt[3]{s}| \le |x_n - \sqrt[3]{s}|^2$, and consequently $|x_n - \sqrt[3]{s}| \le \delta^{2^n}$.

2.17 Show that the value of δ in Exercise 2.16 is $[(\frac{2}{9})\sqrt{21} - \frac{6}{7}]$. (Hint: see inequality (2.70) and its graphical meaning.)

2.18 In analogy to the procedure that leads from (2.48) to (2.52), show how (2.73) follows from (2.72).

2.19 In conjunction with Exercise 2.18, show that the iterations (2.73) satisfy $|x_{n+1} - \sqrt[3]{s}| \le |x_n - \sqrt[3]{s}|^3$.

2.20 Write a computer program for the computation of $\sqrt[3]{s}$, using the second- and third-order methods described in Section 2.7, with both simple and improved initializations. Run this program, using the values $s = 1.728$, 3.375, 5.000, 5.832, and 6.859.

2.21 Show that the second-order methods (for $\sqrt{s}$ and for $\sqrt[3]{s}$) described in Sections 2.4–2.7 have the property that the approximations $x_1, x_2, x_3, \ldots$ are from above.

2.22 Show that for third-order methods (computing $\sqrt{s}$ and $\sqrt[3]{s}$), all the approximations x_n are from above or from below according to whether the initialization x_0 is from above or from below.

2.23 Study (2.10) and (2.48) to develop a second-order iterative formula for the computation of $\sqrt[5]{s}$. (Hint: use four parameters.) Explain why in this case the relevant values of s satisfy $1 \le s < 32$.

2.24 Show that if we use the simple lower initialization (2.86) in the polynomial iterations (2.78), then seven iterations guarantee 15 correct figures.

2.25 Determine the coefficients α, β, and γ in (2.87) so that the resulting polynomial iterative formula for $\sqrt{s}$ is of third order. Show that the values of α, β, and γ obtained are unique.

2.26 Construct second-order polynomial iterations for $\sqrt[3]{s}$, $1 \leq s < 8$, starting from the equality

$$x_{n+1} - \sqrt[3]{s} = (\alpha x_n^2 + \beta x_n + \gamma)(x_n - \sqrt[3]{s})^2$$

Explain why two coefficients are not enough. You should reach the formula $x_{n+1} = x_n[4/3 - x_n^3/(3s)]$.

2.27 Assuming the use of the improved lower initialization $x_0 = (s+6)/7$ for the computation of $\sqrt[3]{s}$ via the iterations obtained in Exercise 2.26, show that $|x_{n+1} - \sqrt[3]{s}| \leq 2|x_n - \sqrt[3]{s}|^2$. Moreover, show that $|x_n - \sqrt[3]{s}| \leq (\frac{1}{2})(2\delta)^{2^n}$, where $0 \leq \sqrt[3]{s} - x_0 \leq \delta$.

2.28 Write a computer program for the computation of $\sqrt{s}$, using the second- and third-order polynomial iterations. Now choose the improved lower initialization and run the program with $s = 1.44$, 1.69, 3.00, and 3.24.

2.29 Repeat Exercise 2.28, this time for the computation of $\sqrt[3]{s}$, using the values $s = 1.728$, 3.375, 5.000, 5.832, and 6.859. (Use the appropriate improved lower initialization.)

2.30 Apply the extended Newton method (2.102) to the function $f(x) = x^2 - s$, and obtain the third-order iterations $x_{n+1} = 3x_n/8 + 3s/(4x_n) - s^2/(8x_n^3)$ for the computation of $\sqrt{s}$. Moreover, show that

$$x_{n+1} - \sqrt{s} = \frac{3x_n + \sqrt{s}}{8x_n^3}(x_n - \sqrt{s})^3$$

and deduce that an upper initialization will lead to

$$|x_{n+1} - \sqrt{s}| \leq \frac{1}{2}\,|x_n - \sqrt{s}|^3$$

2.31 Apply both the Newton–Raphson method (2.100) and the extended Newton method (2.102) to the function $f(x) = 1 - s/x^2$, and obtain the polynomial iterations (2.78) and (2.89), respectively.

2.32 Apply the Newton–Raphson method (2.100) to the function $f(x) = x^3 - s$, and obtain the iterations (2.52).

2.33 Apply both the Newton–Raphson method (2.100) and the extended Newton method (2.102) to the function $f(x) = 1 - s/x^k$, and obtain polynomial iterations for the computation of $\sqrt[k]{s}$. Compare your results for $k = 3$ with Exercise 2.26.

3
Area approximations

3.1 Introduction

Our second numerical laboratory assignment is the computation of an area under a given curve to a desired accuracy. Unlike the calculation of the areas of various polygons, the computation of the area of a circle, an ellipse, or the area under the curve $y = 1/\log x$ from $x = 2$ to $x = 7$, say, is not at all trivial and requires methods of integral calculus or numerical approximations. Such areas arise not only in a geometrical context but also in various applications in engineering, biology, and statistics. In keeping with our policy of making the material as accessible as possible to precalculus students, without sacrificing rigor, we shall somewhat limit the generality so that results can be proved by elementary means and generalizations pointed out.

We shall henceforth be interested in the computation of the area under the graph of a positive function $y = f(x)$, from $x = a$ to $x = b$, but limit ourselves for the time being to monotonic (increasing or decreasing) or convex functions.

3.2 Rectangular approximations

Although the ensuing analysis is carried out in terms of a *general, positive, monotonic* function $f(x)$, $a \leq x \leq b$, it is advisable that the laboratory participants bear in mind a concrete example such as $f(x) = 1/x$, $1 \leq x \leq 2$.

Let us partition the relevant interval $[a, b]$ into n parts of width $h = (b - a)/n$, using the points $x_0 = a$, $x_1 = a + h, \ldots, x_n = a + nh = b$. Next we construct "upper" and "lower" rectangular strips as shown in Figs. 3.1 and 3.2, where $f(x)$ is decreasing. We observe that each upper rectangle in Fig. 3.1 coincides with the curve at its top left corner

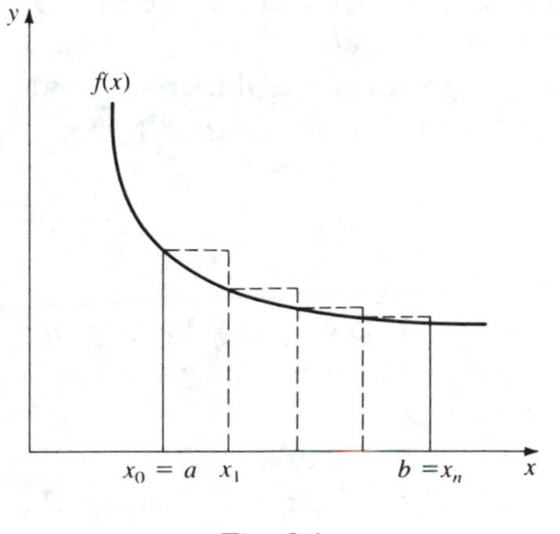

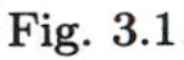

Fig. 3.1

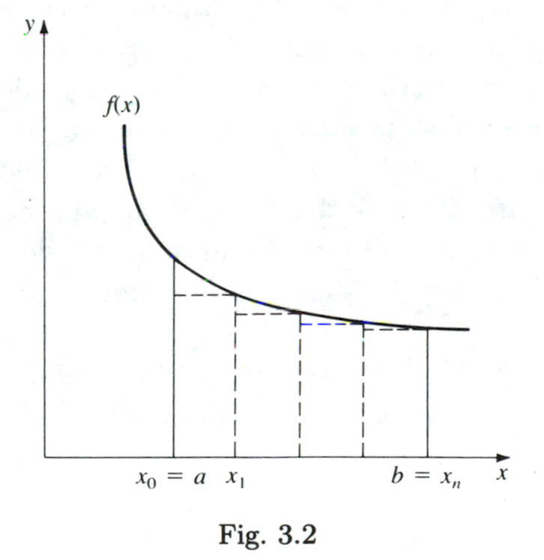

Fig. 3.2

whereas each lower rectangle in Fig. 3.2 coincides with the curve at its top right corner. Accordingly, we denote the sum of the areas of the former by L_n and the sum of the areas of the latter by R_n.

Denoting the required area under the graph by S, we find

$$L_n = \sum_{j=1}^{n} h\, f(x_{j-1}) \geq S \geq \sum_{j=1}^{n} h\, f(x_j) = R_n \tag{3.1}$$

where

$$x_j = a + jh, \qquad x_0 = a, \qquad x_n = b$$

Clearly the inequalities (3.1) should be reversed when $f(x)$ is increasing. In either case, we can trap the required area S between two easily computable sums. Moreover, the difference between those two sums constitutes a bound for the error incurred by using either sum as an approximation to S. This difference B is given by

$$B = |f(b) - f(a)|\frac{b-a}{n} \tag{3.2}$$

and it can be made as small as we please by choosing a sufficiently large n. Accordingly, the error of our approximation can be made to conform to any preassigned tolerance.

To see the geometric interpretation of the error bound (3.2) vividly, consider the next two figures. Figure 3.3 shows the upper and lower rectangular strips, in which each shaded area represents the *local* error bound per strip. If we now shift these shaded areas all the way to the right, we obtain the rectangular shaded column shown in Fig. 3.4, representing the *global* error bound (3.2). Moreover, if we refine the partition in Fig. 3.3, the associated shaded column in Fig. 3.4 will clearly retain its height $|f(b)-f(a)|$ but will possess a correspondingly narrower base. The laboratory instructor might urge the students to build a teaching aid, dubbed *Integroboard*, in which the shaded areas can be physically moved to form the error-bound column. We have used a decreasing function, such as $f(x) = 1/x$, but an increasing function gives the same results (see Exercise 3.1).

Various teaching aids (the preceding is but one example) placed in the mathematical laboratory contribute to the "laboratory atmosphere" that we advocate.

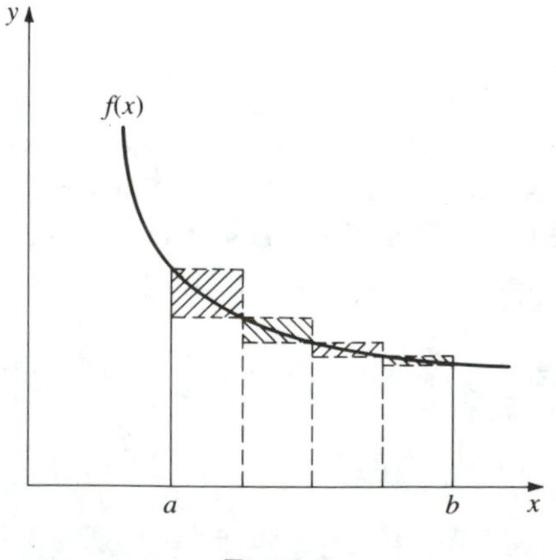

Fig. 3.3

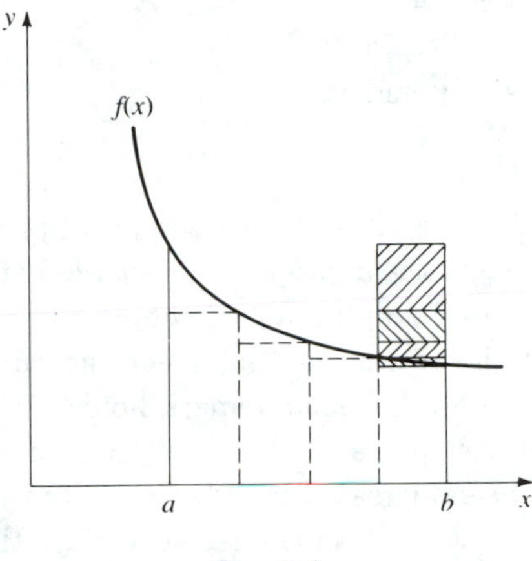

Fig. 3.4

We summarize our results as follows:
Given a function $f(x)$ that is positive and monotonic on $[a,b]$, if we approximate the area under its graph by either

$$\sum_{j=1}^{n} h\, f(x_j) \qquad \text{or} \qquad \sum_{j=1}^{n} h\, f(x_{j-1})$$

then the error incurred will not exceed C/n, where $C = (b-a)|f(b) - f(a)|$.

Next we want to translate our findings into an algorithm, run it on a computer (or a programmable pocket calculator), and obtain numerical results with a prescribed accuracy. Let the given function be $f(x) = 1/x$, and $[a,b] = [1,2]$. If we are interested in q correct decimal figures, we must require that the error should not exceed $\frac{1}{2} \cdot 10^{-q}$. Accordingly, n must satisfy

$$B = (2-1)\left|\frac{1}{2} - 1\right| \frac{1}{n} \leq \frac{1}{2} \cdot 10^{-q} \tag{3.3}$$

so that $n = 10^q$ will do. The following algorithm is suggested:

1. Input $q =$ desired number of correct figures.
2. Set $n = 10^q$ and $h = 1/n$.
3. Set $x = 1$ and $j = 0$.
4. Set $R = L = 0$.
5. Increase j by 1.
6. Replace L by $L + h/x$.
7. Increase x by h.

8. Replace R by $R + h/x$.
9. If $j < n$, return to Step 5.
10. Print the values of L and R.
11. End.

The translation of the algorithm into a computer program must be accompanied, of course, by the appropriate discussion of the various steps, such as why x, R, and L are initialized the way they are. Here L represents the sum of the rectangles that meet the curve at their upper left corner (see Fig. 3.1). A similar remark holds for R. This being the case, we have increased the value of L by h/x prior to Step 7, whereas the value of R has been increased by h/x after Step 7. In the former, x plays the role of x_{j-1} whereas in the latter it plays the role of x_j, after having been increased in Step 7. Although the algorithm is designed to guarantee the required accuracy, we are nevertheless printing both R and L at the end of the calculations to emphasize their coincidence to q decimal figures.

Step 9 merits a few words of explanation. It ensures the end of the calculation when all the rectangles have been added. An alternative method would have been to omit the counter j altogether and to continue the loop as long as $x < 2$ (the end of the interval). This method, however, can lead to false results due to round-off errors. If, for example, $h = \frac{1}{3000}$, the computer calculates this h to finite accuracy only; so when the last rectangle is reached, x is slightly less than 2. Accordingly, the algorithm will call for one additional (but superfluous) rectangle, thus falsifying the results. This is a typical example of a pitfall that would not arise during regular chalk-and-talk sessions.

At this point, it is natural to touch on the subject of *unequal* stripwidths, and related issues. In advanced classes this leads to a fruitful discussion of the concept of area. Moreover, for those who will study calculus, this is a good point of departure for the subjects of definite integrals and numerical integration; but even those who do not study calculus will have learned a good chunk of mathematics, such as upper and lower approximations, error control, and partition refinement in the course of solving a practical problem. In this connection see Exercises 3.2–3.7. Moreover, by printing out various intermediate results, the laboratory participants can actually watch the computational process in the making.

The rectangular-approximation method is rather slow because the associated error is inversely proportional to the number of strips n. Thus, a desired accuracy of six figures necessitates millions of strips. This motivates a search for better methods of area approximation, to which we turn our attention next.

3.3 The trapezoidal method

The preceding approximation used rectangles, which meet the curve at *one point* per strip. It is only natural now to try a set of trapezoids, which meet the curve at *two points* per strip. To maintain the advocated rigorous approach even at the precalculus level, we limit the generality, to start with, to convex (or concave) functions $f(x)$. These concepts should be defined geometrically by using the property that any secant connecting two arbitrary points on such a curve is completely above (or completely below) the curve.

It is good to bear in mind a specific example such as $f(x) = 1/x$ (convex), for $1 \leq x \leq 2$. Alternatively, $f(x) = \sqrt{x}$ (concave) might be used on the same interval.

Now we form the trapezoids depicted in Figs. 3.5 and 3.6. The total sum of their areas is given by

$$T_n = \sum_{j=1}^{n} h \frac{f(x_{j-1}) + f(x_j)}{2}, \qquad x_0 = a, \qquad x_n = b \tag{3.4}$$

where, as before, $h = (b-a)/n$ and $x_j = a + jh$. When $f(x)$ is convex (as in Fig. 3.5), T_n gives an upper approximation for the desired area S.

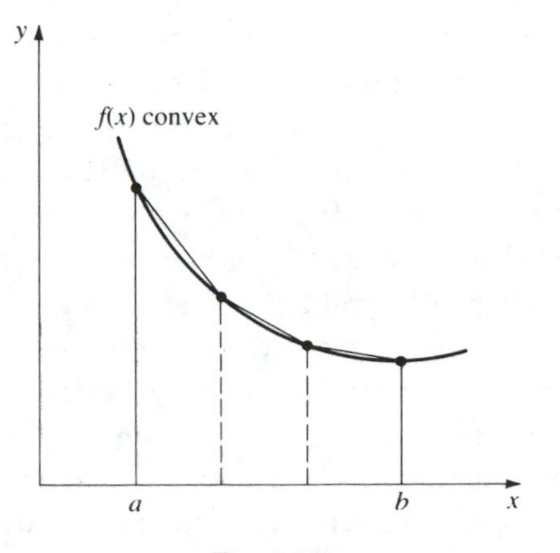

Fig. 3.5

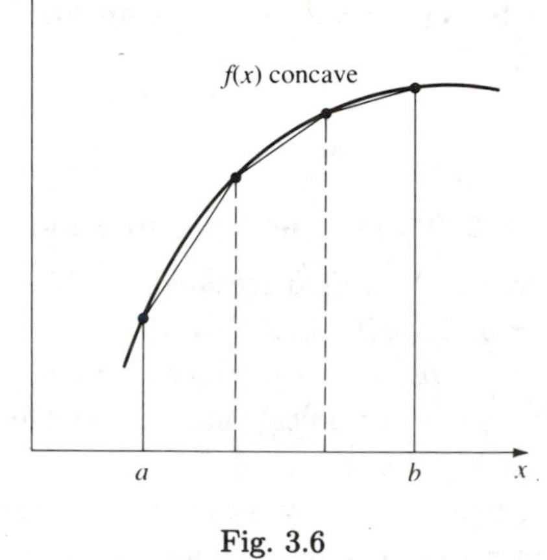

Fig. 3.6

A lower approximation is obtained when $f(x)$ is concave (as in Fig. 3.6). Equation (3.4) can be rewritten as

$$T_n = h\left[\frac{1}{2}f(x_0) + f(x_1) + f(x_2) + \cdots + f(x_{n-1}) + \frac{1}{2}f(x_n)\right] \quad (3.5)$$

and our next objective is to estimate the global error $E = |S - T_n|$ and compare it with the corresponding rectangular error bound (3.2).

To this end, let us draw a concave function $f(x)$ and its associated trapezoids. In addition, we draw an extraneous trapezoid Q, which fits into the extension of $f(x)$ to the right, from $x = b$ to $x = b + h$, as seen in Fig. 3.7. Next, we continue the "roof" of each trapezoid to the left until it covers the adjacent trapezoid (see figure). Had we wanted to extend the roofs to the right, we would have added an extra trapezoid to the left of $x = a$. We now observe that the *local error* (per strip) incurred by the trapezoidal approximation is the difference between the area under the curve $f(x)$ and the area of the corresponding trapezoid. This error, in turn, is seen in Fig. 3.7 to be less than the area of the triangle (such as t_1, t_2, or t_3) that sits on the corresponding trapezoid. If the function $f(x)$ is convex (not concave as in the figure), we would obtain a similar situation except that the roles of the two nonvertical sides of the triangles would be reversed.

Now we take the extreme triangle on the right (t_1 in Fig. 3.7) and slide it leftward until it sits precisely on top of the adjacent triangle (t_2 in Fig. 3.8). Our construction of these triangles enables us to do just that. Next we take t_1 and t_2, together, and slide them leftward until they sit on top of t_3 (see Fig. 3.8). In general, there are n strips, and we carry out

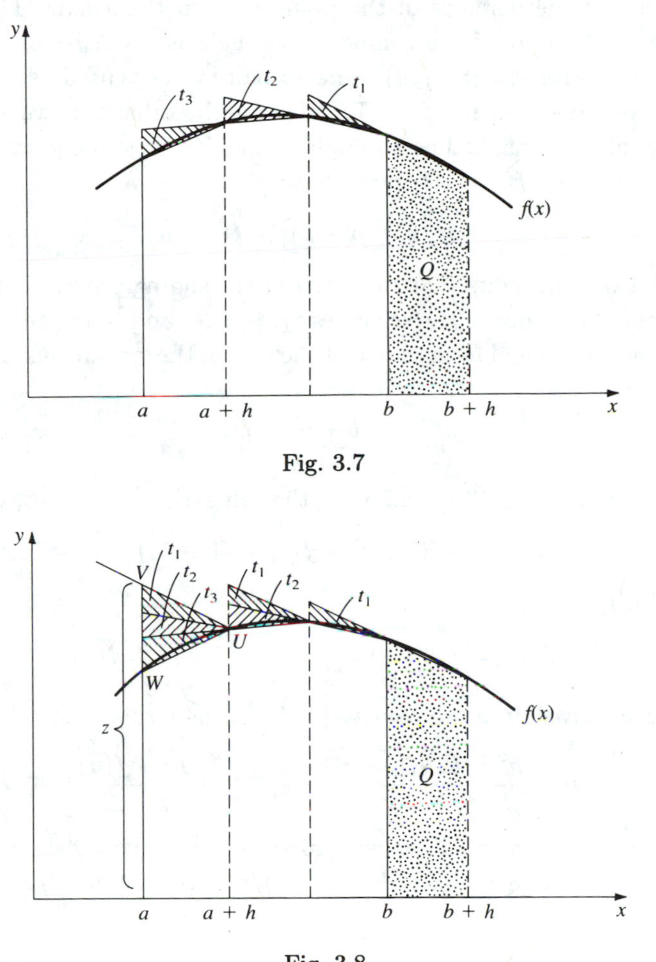

Fig. 3.7

Fig. 3.8

the sliding process just described $(n - 1)$ times. In this way we obtain a composite triangle composed of n little triangles with no overlapping and no holes. The situation for three strips $(n = 3)$ is shown in Fig. 3.8, in which the final, composite triangle has vertices U, V, W. A teaching aid designed to demonstrate this sliding process can be constructed and successfully employed in the mathematical laboratory.

Because the composite triangle consists of the sum of triangles, each of which represents a bound on the local error, its area B gives us a bound for the global error $E = |S - T_n|$. Thus, we have

$$E = |S - T_n| < B = \frac{h}{2}|z - f(a)| \tag{3.6}$$

in which z is the distance of the point V from the x-axis. This is so because the area B of the composite triangle is given by one-half its height h times its base $|z - f(a)|$. The absolute value is used here because for a convex function, $z < f(a)$. To calculate the value of z, we write the equation of the straight line through U and V. This line goes through the point $(a + h, f(a + h))$, so we have

$$y = m[x - (a + h)] + f(a + h) \tag{3.7}$$

A little reflection shows that the effect of the sliding process is to cause the slopes of the tops of the triangles t_1, $t_1 + t_2$, and so on, to be equal; indeed, they are equal to the slope of the roof of the extraneous trapezoid Q. Consequently,

$$m = \frac{f(b + h) - f(b)}{h} \tag{3.8}$$

Substituting $x = a$ in (3.7) and using the value of m from (3.8), we reach

$$z = -f(b + h) + f(b) + f(a + h) \tag{3.9}$$

from which

$$B = \frac{h}{2} |- f(b + h) + f(b) + f(a + h) - f(a)| \tag{3.10}$$

For future convenience, we rewrite (3.10) in the form

$$B = \frac{h^2}{2} \left| \frac{f(b + h) - f(b)}{h} - \frac{f(a + h) - f(a)}{h} \right| \tag{3.11}$$

so that inside the absolute-value symbol we have the difference of two slopes of the type (3.8). Using $h = (b - a)/n$ in the factor $h^2/2$ in (3.11), we reach

$$B = \frac{K}{n^2} \tag{3.12}$$

where

$$K = \frac{(b - a)^2}{2} \left| \frac{f(b + h) - f(b)}{h} - \frac{f(a + h) - f(a)}{h} \right| \tag{3.13}$$

We note that K depends on h, and because $h = (b - a)/n$, K is actually a function of n. Our next objective is to determine a constant D, which is an upper bound for K and independent of n (and h), such that

$$E = |S - T_n| < B = \frac{K}{n^2} \le \frac{D}{n^2} \tag{3.14}$$

This shows that the error bound is inversely proportional to n^2.

To determine D for a typical case, we now investigate the nature of K for the function $f(x) = 1/x$, $1 \le x \le 2$. For this case, (3.13) takes the form

$$K = \frac{(2-1)^2}{2}\left|\frac{1}{h}\left(\frac{1}{2+h} - \frac{1}{2}\right) - \frac{1}{h}\left(\frac{1}{1+h} - 1\right)\right|$$

$$= \frac{1}{2}\left|\frac{1}{1+h} - \frac{1}{2(2+h)}\right| < \left(\frac{1}{2}\right)\frac{1}{1+h} \le \frac{1}{2} = D \qquad (3.15)$$

It follows that in this case K does not exceed $\frac{1}{2}$, regardless of the number of strips used in the approximation. The global error, therefore, is bounded by $(1/2)/n^2$. Thus we see that, whereas the rectangle approximation yields an error bound that is inversely proportional to n, the trapezoidal approximation gives an error bound that is inversely proportional to n^2 (at least for $f(x) = 1/x$).

More generally, a method of approximation with an error bound that is inversely proportional to n^p is called a method of order p. Thus the rectangle method is of first order whereas the trapezoidal method seems to be of second order.

In our particular case, $f(x) = 1/x$ for $1 \le x \le 2$, the attainment of q correct figures requires that $(1/2)/n^2 \le (1/2)10^{-q}$; so $n = 10^{q/2}$ will do. Thus, attaining 4 correct figures necessitates 100 strips, in contradistinction to the 10,000 strips that were needed when the rectangle method was used for the same purpose [see (3.3)]. More generally, we see that the use of a second-order method saves a lot of computational effort (for a given accuracy) because the accompanying error decreases at a faster rate with refinement of the partition (increasing n).

For $f(x) = 1/x$ we obtained $K \le \frac{1}{2}$ independently of n where K, as given by (3.13), is the numerator of the error bound (3.12). In general, for a given convex or concave $f(x)$, it is only necessary to analyze (3.13) and determine a constant D such that $K \le D$ on the underlying interval for all n. Except for the multiplying factor $(b-a)^2/2$, the right-hand side of (3.13) is just the difference of the slopes of the secants associated with the interval endpoints. The only situation that might present a difficulty is the case of a function $f(x)$ possessing vertical (i.e., unbounded) slopes near one or both endpoints of the interval, such as $\sqrt{1-x^2}$ near $x = 1$. If we agree to exclude such cases from our considerations, we can sum up our findings as follows:

Given a function $f(x)$ that is positive and convex (or concave) on $[a,b]$, if we approximate the area S under the curve by T_n given by (3.5), then the error incurred $E = |S - T_n|$ will not exceed D/n^2, where D is a constant independent of n (and h), satisfying

$$D \geq K = \frac{(b-a)^2}{2}\left|\frac{f(b+h)-f(b)}{h} - \frac{f(a+h)-f(a)}{h}\right| \tag{3.16}$$

To gain a better insight into the process of finding $D \geq K$, we shall examine additional examples. Consider the problem of finding the area under the graph of the convex function $f(x) = 1/x^2$, from $x = 2$ to $x = 5$. From (3.16) we find

$$\begin{aligned} K &= \frac{(5-2)^2}{2}\left|\frac{1}{h}\left(\frac{1}{(5+h)^2} - \frac{1}{5^2}\right) - \frac{1}{h}\left(\frac{1}{(2+h)^2} - \frac{1}{2^2}\right)\right| \\ &= \frac{9}{2}\left|\frac{4+h}{4(2+h)^2} - \frac{10+h}{25(5+h)^2}\right| \\ &< \left(\frac{9}{2}\right)\frac{4+h}{4(2+h)^2} < \left(\frac{9}{2}\right)\frac{4+h}{16} \end{aligned} \tag{3.17}$$

where the last inequality follows from the fact that h is positive, and thus the replacement of $4(2+h)^2$ by 16 decreases the denominator and hence increases the fraction. Now, beccause our underlying interval has length $5 - 2 = 3$, it follows that $h \leq 3$. Thus,

$$K < \left(\frac{9}{2}\right)\frac{4+h}{16} \leq \frac{63}{32} = D \tag{3.18}$$

We were somewhat "generous" in obtaining D via (3.17) and (3.18) and could easily have obtained a smaller value – that is, a sharper bound. However, because the essential feature of the error bound D/n^2 is its denominator n^2, we can afford a slightly larger D.

Consider next the concave function $f(x) = \sqrt{x}$, $1 \leq x \leq 4$. This time (3.16) yields

$$K = \frac{(4-1)^2}{2}\left|\frac{\sqrt{4+h}-\sqrt{4}}{h} - \frac{\sqrt{1+h}-\sqrt{1}}{h}\right| \tag{3.19}$$

Multiplying numerators and denominators by the appropriate rationalizing factors, we obtain

$$\begin{aligned} K &= \frac{9}{2}\left|\frac{1}{\sqrt{4+h}+2} - \frac{1}{\sqrt{1+h}+1}\right| \\ &= \frac{9}{2}\left[\frac{1}{\sqrt{1+h}+1} - \frac{1}{\sqrt{4+h}+2}\right] \\ &< \left(\frac{9}{2}\right)\frac{1}{\sqrt{1+h}+1} \le \frac{9}{4} = D \end{aligned} \tag{3.20}$$

The global error $|S - T_n|$ is therefore bounded by $(9/4)/n^2$ in this case, showing again the trapezoidal method to be of second order. Further examples of estimating the global error $|S - T_n|$, as well as determining the number of strips required for a prescribed accuracy, are given in Exercises 3.8–3.10.

To bring the comparison of the rectangular and trapezoidal methods into sharp focus, we suggest setting up the following "competition" in the mathematical laboratory. Let the instructor divide the laboratory participants into two groups, the "Rects" and the "Traps." Each pair of Rects writes a program that computes, say, the area under the curve $f(x) = 1/x$ for $1 \le x \le 2$, with a prescribed accuracy, using the rectangle method. Each pair of Traps does the same using the trapezoidal method. Now all the programs are run simultaneously, and running times are measured. Each pair of students is to stand up as soon as they obtain their final answer. It is clear that the Traps will stand up first (within a few seconds of each other, due to differences among programs) whereas the Rects will rise considerably later. The same experiment can be carried out using different functions, and thus the laboratory participants will witness the relative efficiency of the two methods. This experience will stay with them.

3.4 The midpoint rule

We now present an alternative second-order method for the computation of areas. As before, we limit ourselves to areas under the graphs of convex (or concave) functions. We begin our treatment with a convex function $f(x)$, and the laboratory participants are urged again to bear in mind a specific example, such as $f(x) = 1/x$.

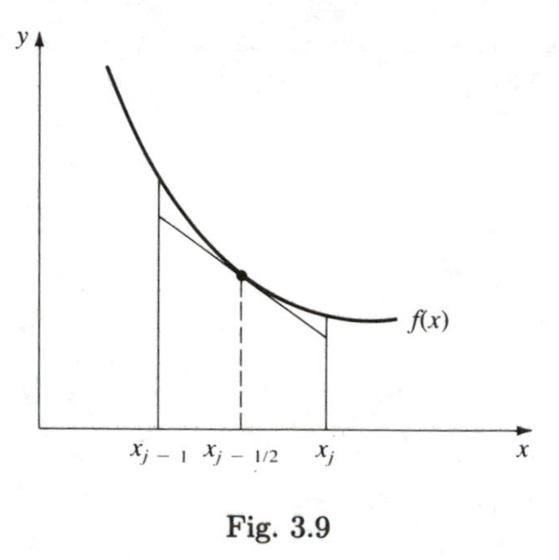

Fig. 3.9

Let us form the so-called midpoint trapezoids, a typical one of which is depicted in Fig. 3.9. The roof of such a trapezoid coincides with the tangent to the graph of $f(x)$ at $x_{j-1/2} = (x_{j-1} + x_j)/2 = a + (j - \frac{1}{2})h$, that is, at the *midpoint* of the interval $[x_{j-1}, x_j]$. The area S under the graph of $f(x)$, from $x = a$ to $x = b$, is now approximated by the sum M_n of the areas of these midpoint trapezoids,

$$M_n = \sum_{j=1}^{n} hf(x_{j-1/2}) \tag{3.21}$$

This is so, because $f(x_{j-1/2})$ is the average of the bases of the jth midpoint trapezoid, h being its height.

To gain insight into the quality of the midpoint-rule approximation to the area S, we first examine a typical strip, such as the jth strip shown in Fig. 3.9. Because $f(x)$ is convex, the area under its graph in the jth strip is larger than the area of the lower (midpoint) trapezoid, $hf(x_{j-1/2})$, but smaller than the area of the upper trapezoid, $h\left[f(x_{j-1}) + f(x_j)\right]/2$, as shown in Fig. 3.10.

Clearly, the geometrical situation is reversed when $f(x)$ is concave. In *both* cases, the local error e_j (per strip), incurred by using either the lower trapezoid or the upper trapezoid, can be bounded in the form

$$e_j \leq h\left|\frac{f(x_{j-1}) + f(x_j)}{2} - f(x_{j-1/2})\right| \tag{3.22}$$

An estimate of the global error $E = \sum e_j$ necessitates the study of (3.22) for each particular $f(x)$ in the underlying interval $[a, b]$. It is instructive

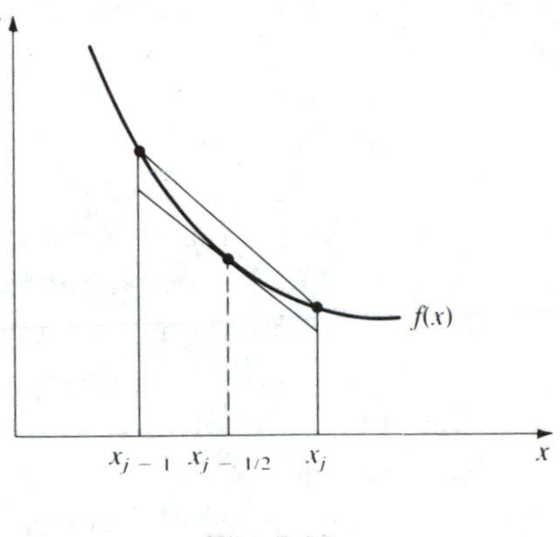

Fig. 3.10

for the laboratory participants to proceed by reexamining $f(x) = 1/x$, $1 \leq x \leq 2$. For this case, (3.22) takes the form

$$e_j \leq \frac{h}{2}\left|\frac{1}{x_{j-1}} + \frac{1}{x_j} - \frac{2}{x_{j-1/2}}\right| \tag{3.23}$$

To simplify the notation, let us write $x_{j-1/2} = u$, so that $x_{j-1} = u - h/2$ and $x_j = u + h/2$. Accordingly,

$$e_j \leq \frac{h}{2}\left|\frac{1}{u+\dfrac{h}{2}} + \frac{1}{u-\dfrac{h}{2}} - \frac{2}{u}\right| = \frac{h^3}{4\left(u-\dfrac{h}{2}\right)u\left(u+\dfrac{h}{2}\right)} \tag{3.24}$$

Clearly $(u - h/2)u(u + h/2) = x_{j-1} \cdot x_{j-1/2} \cdot x_j \geq x_{j-1}^3 \geq 1$, since x_{j-1} is the left endpoint of the base of the jth strip and hence cannot be less than unity in the case under consideration. It follows that

$$e_j \leq \frac{h^3}{4}, \qquad j = 1, 2, \ldots, n \tag{3.25}$$

which is uniformly valid for every strip. Thus, we have

$$|S - M_n| = E = \sum_{j=1}^{n} e_j \leq n \cdot \frac{h^3}{4} = \frac{1}{4n^2} \tag{3.26}$$

since $h = (b-a)/n = (2-1)/n = 1/n$ in our case. Now, the sum of the upper trapezoids is none other then the sum T_n of (3.4), developed for the trapezoidal method in the previous section. Therefore (see Exercise 3.11), we also have

$$|S - T_n| \leq \frac{1}{4n^2} \tag{3.27}$$

The instructor should point out to the laboratory participants that when an analysis similar to the preceding is carried out to obtain a bound on $|S - M_n|$ for any convex (concave) function, this same bound will also do for $|S - T_n|$. This provides us with an alternative way of estimating $|S - T_n|$, which is different from (3.16). In this connection see Exercise 3.12.

It is clear, geometrically, that for both convex and concave functions $f(x)$, the required area S is sandwiched between M_n and T_n (see again Exercise 3.11). Accordingly, if for a given $f(x)$ it is difficult to derive practical estimates based on (3.22), which lead to a priori error bounds of the type (3.26), there is always the following alternative. Let the laboratory participants compute T_n and M_n, for some initial n. Because both $|S - T_n|$ and $|S - M_n|$ are bounded by $|T_n - M_n|$, it follows that whenever $|T_n - M_n| \leq \frac{1}{2} \cdot 10^{-q}$, they have actually computed S to q (prescribed) correct figures. If, however, $|T_n - M_n| > \frac{1}{2} \cdot 10^{-q}$, let them restart the computations with n replaced, say, by $2n$. In this connection see Exercise 3.13. The instructor might well point out that when an error-bound analysis, leading to an estimate of the type (3.26), can actually be carried out, it is not only preferable but it enables us to determine, a priori, the number n of strips required to ensure a prescribed accuracy.

It is instructive to examine another example, compute and bound the local errors, see how they accumulate to produce the global error, and subsequently find a practical overall error bound. Thus, consider the area S under the graph of $f(x) = x^3$, from $x = 0$ to $x = 2$. For this case, (3.22) shows that the local error e_j satisfies

$$e_j \leq \frac{h}{2}\left|\left(u - \frac{h}{2}\right)^3 + \left(u + \frac{h}{2}\right)^3 - 2u^3\right| = \frac{3}{4}uh^3 \tag{3.28}$$

where $u = x_{j-1/2}$. Now $u \leq 2$ for each strip in the underlying interval, so we have

$$e_j \leq \frac{3}{2}h^3 = \frac{12}{n^3} \tag{3.29}$$

since $h = (b - a)/n = 2/n$. Note that the local error e_j is bounded by a quantity inversely proportional to n^3, as has been the case for

$f(x) = 1/x$, $1 \le x \le 2$, shown in (3.25). The local error estimate (3.29) leads immediately to the global error bound

$$|S - M_n| = E = \sum_{j=1}^{n} e_j \le n \cdot \frac{12}{n^3} = \frac{12}{n^2} \tag{3.30}$$

The global error is seen to be bounded by a quantity inversely proportional to n^2, as was the case in (3.26). This is a general phenomenon and shows that the accumulation of local errors causes the global error to increase by a factor proportional to n. Moreover, if we want the error in (3.30) to be less than or equal to $\frac{1}{2} \cdot 10^{-q}$, we readily find that the number n of strips must merely satisfy $n \ge \sqrt{24} \cdot 10^{q/2}$. An analogous statement applies to (3.26) for the case $f(x) = 1/x$. We remark that the global error bound (3.30) can be sharpened by using (3.28) directly, rather than (3.29). In this connection see Exercise 3.14.

The midpoint rule and the trapezoidal method have been shown to be of second order. Thus, they are superior to the first-order rectangular method. Indeed, for a prescribed accuracy of q correct figures, the number n of strips required for the rectangular method is proportional to 10^q, whereas for the second-order methods n need only be proportion to $10^{q/2}$ – which is evidently much more efficient. Additional examples demonstrating this efficiency can be found in Exercises 3.16 and 3.17. If the function $f(x)$ under consideration is, say, concave for $a \le x \le c$ and convex for $c \le x \le b$, the area under its graph from $x = a$ to $x = b$ can be computed by treating these two subintervals separately, as in Exercise 3.18.

3.5 A more advanced viewpoint

This section on area-approximation methods, viewed as numerical integration and using calculus methods, may be omitted without loss of continuity.

In accordance with our general ideas of presenting material for the numerical laboratory, we have employed a precalculus approach to the computation of areas. However, when combined later with calculus, the methods and error estimates can be generalized to cover a much wider class of functions. Moreover, the entire analysis just given can serve as an excellent introduction to the study of integral calculus and numerical integration. The computation of certain areas also becomes more meaningful when viewed from a calculus angle. For example, when we want to compute the value of $\ln c$ for a given positive c, we can write

$$c = b \cdot 2^k, \qquad 1 \le b < 2, \quad k \text{ integer} \tag{3.31}$$

so that

$$c = \ln b + k \ln 2 \tag{3.32}$$

Thus, the computation of the area under the graph of $f(x) = 1/x$ for $a = 1 \le x \le b$ (which actually equals $\ln b$) turns out to be much more than meets the eye at the precalculus level. Indeed, once we compute $\ln 2$, we can use (3.32) to calculate $\ln c$ for any positive c by merely finding $\ln b$, where $1 \le b < 2$ is given in (3.31). This and similar interpretations can be hinted at by the laboratory instructor even at the precalculus level.

When we want to compute the value of the integral

$$I = \int_a^b f(x)dx \tag{3.33}$$

where the integrand $f(x)$ is any differentiable or twice-differentiable function in $[a, b]$, the rectangular, trapezoidal, and midpoint methods are still applicable, as can be shown with calculus. Moreover, if we set

$$M_1 = \operatorname*{Max}_{a \le x \le b} |f'(x)|, \qquad M_2 = \operatorname*{Max}_{a \le x \le b} |f''(x)| \tag{3.34}$$

we find that the global errors incurred by those three methods satisfy

$$|I - L_n| \le \frac{C}{n}, \quad |I - R_n| \le \frac{C}{n}, \qquad C = \frac{M_1}{2}(b-a)^2 \tag{3.35}$$

$$|I - T_n| \le \frac{D}{n^2}, \qquad D = \frac{M_2}{12}(b-a)^3 \tag{3.36}$$

$$|I - M_n| \le \frac{H}{n^2}, \qquad H = \frac{M_2}{24}(b-a)^3 \tag{3.37}$$

These error bounds constitute generalizations of the results obtained by precalculus methods in the previous three sections. These results apply to a wide class of functions, not necessarily monotone or convex (concave). In this connection see Exercises 3.19 and 3.20.

At this point, further improvement of the approximation methods can be considered. We have seen that the rectangle method employs just one point, locally, of the graph of $f(x)$ and that the trapezoidal method, which uses two points locally, leads to a considerable improvement. This being the case, it is only natural to seek methods yielding higher-order approximations through the use of three (or more) points locally. This idea leads to Simpson's rule and various other advanced methods of numerical integration.

Looking back at the material presented in this chapter (except this last section), the laboratory participants will find that we have actually laid the foundations for the concepts of the definite integral and numerical integration, using only precalculus means, accompanied by detailed computations in the numerical laboratory.

Exercises

3.1 Given an increasing, positive function $f(x)$, $a \le x \le b$, approximate the area S under its graph, using the sum L_n of left rectangles as well as the sum R_n of right rectangles. Observe that the global error-bound column will be positioned at the left end of the interval $[a, b]$, compute its area, and show that it is equal to B, as defined in (3.2).

3.2 Write and run a program that computes the area under the graph of $f(x) = 1/x$, from $x = 1$ to $x = 2$, using the rectangular approximation and ensuring four correct figures. Repeat the process, requiring five correct figures (observe that you have obtained approximations for $\ln 2$).

3.3 Carry out Exercise 3.2 for $f(x) = x^2$, $0 \le x \le 3$.

3.4 Compute again the area under the graph of $f(x) = x^2$, $0 \le x \le b$, but this time find R_n and L_n analytically, using the formula $\sum_{j=1}^{n} j^2 = n(n+1)(2n+1)/6$. Show that

$$\frac{b^3}{3}\left(1-\frac{1}{n}\right)\left(1-\frac{1}{2n}\right) \le S \le \frac{b^3}{3}\left(1+\frac{1}{n}\right)\left(1+\frac{1}{2n}\right)$$

and compare with Exercise 3.3, using $b = 3$ and the values of n that were determined there. (This method was introduced by Archimedes to find the area under a parabola, about 2000 years before the development of calculus.)

3.5 Using the rectangular method, write and run a program that computes the area under the graph of $f(x) = \sqrt{1-x^2}$ from $x = 0$ to $x = 1$, to four correct figures. Note that $S = \pi/4$.

3.6 Repeat Exercise 3.5 for $f(x) = 1/(1+x^2)$, $0 \le x \le 1$. (Using calculus, it can be shown that in this case, too, $S = \pi/4$.)

3.7 We are given a positive function $f(x)$ that increases from $x = a$ to $x = c$ and decreases from $x = c$ to $x = b$ and are interested in finding the area S under its graph from $x = a$ to $x = b$. Explain how to use the rectangular method in each subinterval separately, so that S can be "sandwiched" between appropriate lower and upper approximations.

3.8 Suppose we are interested in computing the area under the graph of $f(x) = x^2$, $0 \le x \le 3$, using the trapezoidal method. First find K of (3.16), and observe that in this case you can take $D = K$. Explain why. Next determine the number n of strips required to ensure four and five correct figures, respectively, and compare these values of n with the corresponding values in Exercise 3.3. Write and run an appropriate program.

3.9 Compute the area under the graph of $f(x) = x^3$, $\frac{1}{2} \le x \le \frac{3}{2}$, using both the trapezoidal and rectangular methods. Determine the corresponding constants that arise in the error bounds, and the associated number n of strips required to ensure four correct figures in each case. Write and run appropriate programs and compare their running times.

3.10 Determine the number n of strips required to ensure five correct figures when the trapezoidal method is used to compute the area under the graph of $f(x) = \sqrt{1 + x^2}$, $1 \le x \le 3$. It can be shown that, in this case,

$$\frac{f(3+h) - f(3)}{h} > \frac{f(1+h) - f(1)}{h} > 0$$

because the graph is much steeper near $x = 3$ than near $x = 1$. To see this, draw the graph. This being the case, you can simplify the analysis by using the inequality

$$\frac{f(3+h) - f(3)}{h} - \frac{f(1+h) - f(1)}{h} < \frac{f(3+h) - f(3)}{h}$$

A similar argument was employed in (3.20).

3.11 Explain why, for any positive convex (concave) function, we have

$$|S - M_n| \le |T_n - M_n| \qquad \text{and} \qquad |S - T_n| \le |T_n - M_n|$$

3.12 Review the trapezoidal approximation T_n to the area S under the graph of $f(x) = 1/x$, $1 \le x \le 2$, and its global error estimate D/n^2 developed in Section 3.3, leading to (3.15). Also review the alternative error estimate in Section 3.4, leading to (3.27). Explain the different natures of these two estimates with the aid of an enlarged drawing of one typical strip, in which the areas representing the respective local error bounds are colored differently.

3.13 Suppose we want to compute the area S under the graph of the concave function $f(x) = \sqrt{16 - x^8}$, from $x = 0$ to $x = 1$. Write and run a program that calculates M_n and T_n for some initial n, say $n = 10$. If $|T_n - M_n| > \frac{1}{2} \cdot 10^{-5}$, the program should restart the calculation with n replaced by $2n$ and repeat the process until $|T_n - M_n| \leq \frac{1}{2} \cdot 10^{-5}$, ensuring five correct figures for S. Note that when n is replaced by $2n$, the quantity $|T_n - M_n|$ decreases by a factor of about $\frac{1}{4}$, reflecting the second-order nature of the approximations T_n and M_n.

3.14 Reconsider the computation of the area S under the graph of $f(x) = x^3$, $0 \leq x \leq 2$, discussed in the text. Instead of passing "generously" from (3.28) to (3.29), proceed from (3.28) to compute

$$\sum_{j=1}^{n} e_j \leq \sum_{j=1}^{n} \frac{3}{4} \left[\left(j - \frac{1}{2} \right) h \right] h^3$$

which involves the sum of an arithmetic progression. Finally, show that the resulting bound is sharper than (3.30).

3.15 Repeat Exercise 3.14 for a general interval $a \leq x \leq b$, $a \geq 0$. Show that the resulting error bound is given by the quantity $\frac{3}{8}(b + a)(b - a)^3/n^2$ whereas the method used in the text, via (3.29), leads to the coarser bound $\frac{3}{4}b(b - a)^3/n^2$.

3.16 Write and run a program that computes the area S under the graph of the convex function $f(x) = 1/x^2$, $2 \leq x \leq 5$, using the midpoint rule M_n. Estimate the local and global errors and determine the number n of strips required to ensure six correct figures. How many strips would have been needed to achieve the same accuracy if the rectangular method had been used?

3.17 Repeat Exercise 3.16 in its entirety for the concave function $f(x) = \sqrt{x}$, $1 \leq x \leq 4$. You will have to rationalize certain expressions, repeatedly, to arrive at the required estimates.

3.18 Consider the area S under the graph of $f(x) = 1/(3+x^2)$, from $x = 0$ to $x = 3$. This function is concave for $0 \leq x \leq 1$ and convex for $1 \leq x \leq 3$. (The point $x = 1$ is a point of transition from concavity to convexity, and is called an inflection point.) Write and run a program that uses the approximations T_n and M_n in each subinterval, separately, and then combines the results appropriately to obtain upper and lower approximations for S with four correct figures.

3.19 Find the various global error bounds, represented in (3.35)–(3.37), for $f(x) = 1/(3 + x^2)$, $0 \le x \le 3$. Compare these with the results of Exercise 3.18. Note that the use of calculus avoids considerations of subintervals in which $f(x)$ is convex or concave.

3.20 Given $f(x) = x^2 e^{-x^2}$ in the interval $0 \le x \le 4$, we want to find the area S under its graph. Use the bounds given by (3.35)–(3.37) to determine the number n of strips required by the various approximation methods to obtain the value of S correct to q figures. Observe that $f(x)$ is neither monotone nor convex nor concave, over the entire given interval. Moreover, the indefinite integral of $f(x)$ is not expressible in terms of elementary functions.

4
Linear systems – An algorithmic approach

4.1 Introduction

In this chapter we present an algorithmic approach to the solution of systems of linear equations, another typical subject for the mathematical laboratory. No knowledge of matrices, vectors, and their underlying theory is presupposed, and thus the laboratory participants can handle this material even before the study of linear algebra.

After the development of an algorithm for the solution of "naive" systems of linear equations, special attention will be paid to problematic cases in which unrealistic answers with huge errors might be obtained. In particular, we shall discuss reasons for loss of accuracy, sensitivity to minor changes in the data, pivoting, scaling, and computational efficiency. By elaborating on each of these points by means of appropriate examples, we hope to present this traditionally abstract mathematical subject in a concrete, practical way that will be more meaningful to many students.

4.2 Coefficient tables

Systems of linear equations arise naturally in many practical areas such as mixing liquids, work and power calculations, electrical circuit computations, and marketing problems. It is particularly useful to demonstrate the subject under consideration by means of 3×3 systems (three equations and three unknowns). Such systems are not too large and cumbersome, but nevertheless constitute a case in which a pattern is revealed. Occasionally, when it is necessary for clarity, 4×4 and 2×2 systems will also be used.

We start with the following example: A pharmacist buys 100 units of substance α, 164 units of substance β, and 27 units of substance γ.

Using these substances, she prepares three medicines A, B, and C, each of which contains the substances α, β, and γ according to the following table:

	A	B	C
α	10%	30%	50%
β	40%	20%	30%
γ	2%	9%	15%

(4.1)

Obviously, the medicines contain other components such as distilled water. If the laboratory instructor prefers, the letters A, B, and C can be replaced by "pills," "capsules," and "ointment." Now, the pharmacist wants to prepare $X(1)$ units of A, $X(2)$ units of B, and $X(3)$ units of C. What should $X(1)$, $X(2)$, and $X(3)$ be so that all the substances bought would be used up completely? We point out that it is preferable to use the notation $X(1), X(2), X(3)$ for the unknowns, rather than x, y, z, so that they will form a sequence and be easy to handle in automatic processing. The system of equations thus obtained is

$$\begin{aligned} 0.10X(1) + 0.30X(2) + 0.50X(3) &= 100 \\ 0.40X(1) + 0.20X(2) + 0.30X(3) &= 164 \\ 0.02X(1) + 0.09X(2) + 0.15X(3) &= \ 27 \end{aligned} \tag{4.2}$$

To make the system more convenient, we perform some algebraic operations on it, such as multiply the first equation by (-4) and add it to the second. When we do this, the coefficient of $X(1)$ in the second equation vanishes (which means that $X(1)$ "disappears" from the second row). Similar operations can be performed at will, and we can see that the net effect is a change in the coefficients on the left-hand side and the numbers on the right-hand side. Because $X(1), X(2), X(3)$ and the equality signs are left unchanged, it suffices to represent the system by the following table:

0.1	0.3	0.5	100
0.4	0.2	0.3	164
0.02	0.09	0.15	27

(4.3)

Every row represents the left-hand side of one of the equations in (4.2) whereas the rightmost column represents the right-hand sides in (4.2). For clarity, we separate this column by a vertical line. The table (4.3), exclusive of the rightmost column, is called the "coefficient table" whereas the full table is the "augmented coefficient table" or the "augmented table." Whenever the number of unknowns equals the number of equations, the coefficient table has an equal number of rows and columns (here 3×3). The augmented table always has one additional (rightmost) column. Associated with this table we suggest forming a column of the unknowns,

$$\boxed{\begin{array}{c} X(1) \\ X(2) \\ X(3) \end{array}} \tag{4.4}$$

At the end of the solving process, which we shalll discuss at length, the components of this column will turn into numbers that represent the desired solution; that is, the column of unknowns will become the "solution column." The solving process can be performed step by step on the augmented table becausse it fully reflects the current state of the system of equations. In particular, multiplying one row by an appropriate factor and adding it to another row is a permissible operation on the table.

The question of how to store the augmented table in the computer's memory should now be raised. As far as the user is concerned, it is natural and convenient to store it in tabular form, that is, in a two dimensional array with rows and columns. Denoting this table by T, we refer to the first entry in the first row as $T(1,1)$, the second entry in the third row as $T(3,2)$, and so on. The first index is the row index, as is usual in most programming languages, and the second index is the column index. The fact that T is a "table variable" (a two-dimensional array) should of course be declared at the beginning of the solving process. The column of unknowns should also be declared initially. It has, of course, only a row index, as in (4.4). Thus, we can rewrite the system in the following form:

$$\begin{aligned} T(1,1) \cdot X(1) + T(1,2) \cdot X(2) + T(1,3) \cdot X(3) &= T(1,4) \\ T(2,1) \cdot X(1) + T(2,2) \cdot X(2) + T(2,3) \cdot X(3) &= T(2,4) \\ T(3,1) \cdot X(1) + T(3,2) \cdot X(2) + T(3,3) \cdot X(3) &= T(3,4) \end{aligned} \tag{4.5}$$

Note that $N(N+1) + N = N^2 + 2N$ memory locations are needed for the augmented table and the column of unknowns, for the case of an $N \times N$ system (in our example $N = 3$).

4.3 Triangularization

In the following we treat a 3×3 linear system using the notation just introduced, but nevertheless we recommend that laboratory participants bear in mind a specific numerical example. Let us assume for the moment that we have a 3×3 system of the particular form

$$\begin{aligned} T(1,1) \cdot X(1) + T(1,2) \cdot X(2) + T(1,3) \cdot X(3) &= T(1,4) \\ T(2,2) \cdot X(2) + T(2,3) \cdot X(3) &= T(2,4) \\ T(3,3) \cdot X(3) &= T(3,4) \end{aligned} \tag{4.6}$$

For this case, the augmented table can be represented symbolically by

$$\begin{array}{|ccc|c|} \# & \# & \# & \# \\ 0 & \# & \# & \# \\ 0 & 0 & \# & \# \end{array} \tag{4.7}$$

in which the #'s represent the coefficients in (4.6), and we see that

$$T(2,1) = T(3,1) = T(3,2) = 0 \tag{4.8}$$

Moreover, the #'s in the coefficient table form a triangle, and hence such a system is said to be in *triangular form*. We are interested in triangular systems because they are easy to solve. Indeed, from the last equation in (4.6) we immediately have

$$X(3) = T(3,4)/T(3,3) \tag{4.9}$$

The laboratory instructor should point out that it is assumed at this stage that the denominator $T(3,3)$, as well as the denominators to be used subsequently, are different from zero. We remove this assumption later. Knowing $X(3)$, we can *back substitute* it in the second equation, obtaining

$$X(2) = \big[T(2,4) - T(2,3) \cdot X(3)\big]/T(2,2) \tag{4.10}$$

Finally, we back substitute $X(3)$ and $X(2)$ in the first equation, arriving at

$$X(1) = \big[T(1,4) - T(1,3) \cdot X(3) - T(1,2) \cdot X(2)\big]/T(1,1) \tag{4.11}$$

Because a linear system is usually not in triangular form, it is natural to look for ways of constructing a *triangularization algorithm* that will transform a given system into an equivalent triangular system. We suggest starting with a specific example such as the pharmacist problem, for which the augmented table is given by (4.3). Triangularization will be achieved via permissible operations: The first row operates on the the rows below it, then the second row operates on the rows below it (in the pharmacist's problem, on the third and last row), and so on, which leads to the desired triangular form. In this connection see Exercises 4.1 and 4.2.

After this triangularization, we apply the back substitution procedure to compute the solution column (see Exercises 4.3–4.6). The entire process, consisting of triangularization and back substitution, is referred to as *Gaussian elimination*, after the great mathematician Karl Friedrich Gauss (1777–1855), who first suggested it.

4.4 Gaussian elimination algorithm

We assume that Exercises 4.1–4.4 have been worked out in detail and now develop a Gaussian elimination algorithm for a general system of N equations and N unknowns ($N \times N$), for which the augmented table T has N rows and $N+1$ columns. The element in the Rth row and the Cth column is denoted by $T(R, C)$. The laboratory participants will recall that $T(R, N+1)$, that is, the $(N+1)$ entry in the Rth row, is the number on the right-hand side of the Rth equation. To start the triangularization process, we operate with the first row on the second one, multiplying the first row by an appropriate factor and adding it to the second so that the resulting row has a vanishing first element, that is, $T(2, 1) = 0$. We operate in like manner with the first row on *all* the rows below it, and then proceed to operate with the second (revised) row on the succeeding rows below it, until the next-to-the-last row operates on the last one. The last row, of course, never functions as an "operating row." The consecutive operations of rows on succeeding rows are controlled by a loop, starting with the first row ($R = 1$), continuing to the second row ($R = 2$), and ending with the $(N - 1)$ row. This loop will be referred to as the operating-row loop. (It is assumed that laboratory participants are familiar with elementary programming concepts such as loops and arrays.)

For each index R of the operating row, the index of the rows being operated on, say K, varies from $R + 1$ to N because these are the indices of the rows following the Rth row. The handling of the rows being operated on at each stage is controlled by a second loop, nested within the operating-row loop, that is called the middle loop.

When the operating row (the Rth row) operates on a row below it (the Kth row), the operation runs from column to column; that is, a third loop takes us from the first column (so it seems at first, but see Exercise 4.7) to column $N+1$. In other words, this innermost loop actually carries out the arithmetic operations of the Gaussian elimination, as follows: It takes us from column to column along the row in question, adding to each term the corresponding term of the operating row multiplied by an appropriate factor F, thus eliminating the term we want to vanish.

The algorithm of the triangularization can be described schematically as follows:

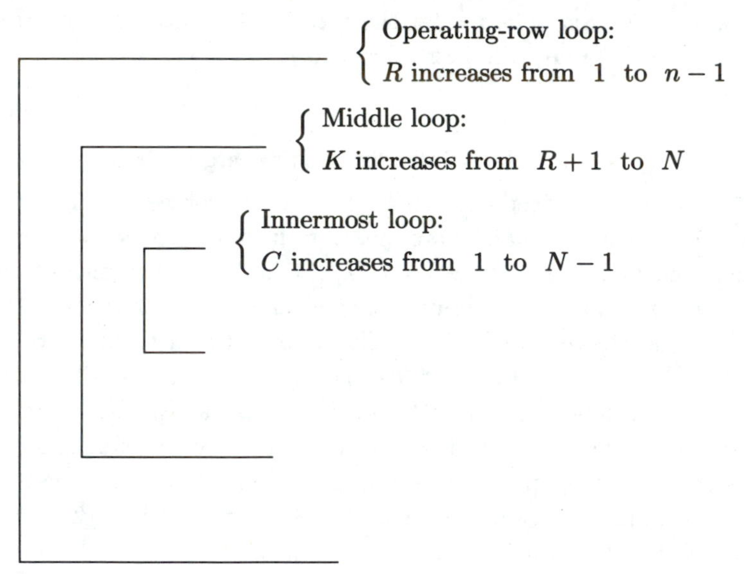

The factor F will be computed in such a way [see (4.14)] that in the coefficient table a new zero is formed at the desired position in the Kth row. Thus, in the innermost loop, we add the quantity $F \cdot T(R,C)$ to $T(K,C)$, thereby generating the new $T(K,C)$. In other words,

$$[T(K,C)]_{\text{new}} = [T(K,C)]_{\text{old}} + F \cdot T(R,C) \tag{4.12}$$

When row R operates on row K, the factor F is the same for all columns, and therefore it should be computed before entering the innermost loop. That is to say, F should be computed in the middle loop.

Note that the creation of new zeros in the coefficient table leaves all previously created zeros unaffected. For precisely this reason, there is no point in operating on all the columns in the Kth row; it is sufficient that the column-loop index C vary from R to $N+1$. In other words, this index C should start at R rather than at 1, thus making the algorithm more efficient (see again Exercise 4.7). The mechanics of the entire triangularization process can best be understood by viewing the augmented table at a typical stage of the triangularization:

$$
\left[\begin{array}{ccccc|c}
T(1,1) & \cdots & \cdots & \cdots & T(1,N) & T(1,N+1) \\
0 & \ddots & & & \vdots & \vdots \\
\vdots & \ddots & & & \vdots & \vdots \\
0 & & 0 & T(R,R) \cdots & T(R,N) & T(R,N+1) \\
\vdots & & \vdots & \vdots & \vdots & \vdots \\
0 & & 0 & T(K,R) \cdots & T(K,N) & T(K,N+1) \\
\vdots & & \vdots & \vdots & \vdots & \vdots \\
0 & \cdots & 0 & T(N,R) \cdots & T(N,N) & T(N,N+1)
\end{array}\right] \tag{4.13}
$$

The next objective is to generate zeros – by means of permissible operations – in all the entries of the Rth column under $T(R,R)$. A little reflection, on setting $C = R$ in (4.12), shows that the appropriate factor F is given by

$$F = -\frac{T(K,R)}{T(R,R)} \tag{4.14}$$

For F to be well defined, we assume for the time being that the so-called pivot element $T(R,R)$ in (4.14) differs from zero and is not even nearly equal to zero. This assumption is made (and later removed) for each pivot element, $T(R,R)$, in all the operating rows. A system for which this assumption holds is referred to as a *naive case.* The expression for F in (4.14) does not depend on C, and thus it should be computed in the middle loop before starting the innermost column loop. This important point is an illustration of a general efficiency principle: Any computation that can be carried out ahead of starting a loop, should not be placed inside it, thus avoiding unnecessary, repetitious calculations.

Before we set up the triangularization algorithm in detail, we must remember to start by declaring T as a table variable (a two-dimensional array) and X as a column variable (a one-dimensional array). The first data item supplied should be the number N of equations, which of course must not exceed the number of rows set aside in the declaration of T. The next data to be supplied are the coefficients of the augmented table – whose actual dimension is $N \times (N+1)$ – which are customarily read row after row. Accordingly, the following *naive triangularization algorithm* is suggested (for, say, $N \le 50$):

1. Declare T as a two-dimensional array, with 50 rows and 51 columns.
2. Declare X as a one-dimensional array, with 50 components.
3. Input the value of N (not exceeding 50).
4. Repeat Steps 4–8 for R increasing from 1 to N.
5. Repeat Steps 5–7 for C increasing from 1 to $N+1$.
6. Input the next data item into $T(R,C)$.
7. Close the loop started in Step 5.
8. Close the loop started in Step 4.
9. Repeat Steps 9–16 for R increasing from 1 to $N-1$.
10. Repeat Steps 10–15 for K increasing from $R+1$ to N.
11. Compute $F = -T(K,R)/T(R,R)$.
12. Repeat Steps 12–14 for C increasing from R to $N+1$.
13. Replace the value of $T(K,C)$ by $\big[T(K,C) + F \cdot T(R,C)\big]$.
14. Close the loop started in Step 12.
15. Close the loop started in Step 10.
16. Close the loop started in Step 9.

Presenting the algorithm in this form paves the way for a straightforward translation into whatever programming language is used in the mathematical laboratory. Clearly, the double loop in Steps 4–8 handles the input of the augmented table whereas the triple loop in Steps 9–16 actually carries out the triangularization.

Because our system is a naive one, by assumption, the preceding algorithm generates a triangular system equivalent to the original one. Thus, our next objective is to obtain the solution by back substitution. Starting with the last unknown, we can immediately write

$$X(N) = T(N,N+1)/T(N,N) \tag{4.15}$$

and then, for values of R decreasing from $N-1$ to 1,

$$X(R) = \Big[T(R,N+1) - \sum_{J=R+1}^{N} T(R,J)X(J)\Big]/T(R,R) \tag{4.16}$$

Accordingly, we may append to the triangularization process the following *back-substitution algorithm*:

17. Compute $X(N) = T(N,N+1)/T(N,N)$.
18. Repeat Steps 18–23 for R decreasing from $N-1$ to 1.
19. Repeat Steps 19–21 for C increasing from $R+1$ to N.
20. Replace $T(R,N+1)$ by $\big[T(R,N+1) - T(R,C) \cdot X(C)\big]$.

21. Close the loop started in Step 19.
22. Compute $X(R) = T(R, N+1)/T(R, R)$.
23. Close the loop started in Step 18.
24. Repeat Steps 24–26 for R increasing from 1 to N.
25. Print out the values of R and $X(R)$.
26. Close the loop started in Step 24.
27. End.

At this point it is suggested that the laboratory participants write and run a program along the lines just indicated for various systems of different orders. It is a good idea to have the program print out the augmented table at several intermediate stages, such as between Steps 15 and 16, to observe the triangularization process in the making. In this connection see Exercises 4.8–4.13.

It turns out that the Gaussian elimination method for the solution of $N \times N$ linear systems is rather efficient. The next section examines the actual computational work involved.

4.5 Computational complexity

We now estimate the computational work associated with the Gaussian elimination, as a function of N, which is its *computational complexity.* We start by examining the three nested loops that triangularize a given system of order N. In this process, the first row operates on the $N-1$ rows below it, the second row operates on the $N-2$ rows below it, and so on, until finally the $(N-1)$ row operates on the last row. Thus, the number of row operations is given by

$$(N-1) + (N-2) + \cdots + 2 + 1 = \frac{N(N-1)}{2} = \frac{1}{2}(N^2 - N) \quad (4.17)$$

Assuming that each of these row operations is performed along the entire row ($N+1$ elements), and denoting by M the number of multiplications and by A the number of additions involved, we find that

$$M = A = (N+1)\frac{(N^2 - N)}{2} = \frac{1}{2}(N^3 - N) \quad (4.18)$$

We call one multiplication and one addition a *computational work unit*, so the triangularization uses about $N^3/2$ units, where for simplicity we neglect the linear term $N/2$, which is much smaller than $N^3/2$. If we take into account the fact that the row operations are not performed along the entire row [see (4.12) and the subsequent paragraph], we can sharpen this estimate and get $N^3/3$. In this connection see Exercise 4.14.

To estimate the computational complexity of the back substitution as well, we note that finding $X(N)$ necessitates one division, which we consider to be one computational unit. Next, $X(N-1)$ can be found by means of one multiplication and one subtraction (constituting one unit) followed by a division, which gives a total of two computational work units. Continuing in this manner, we find that the total number of units needed for the back substitution is given by

$$1+2+3+\cdots+N=\frac{N(N+1)}{2}=\frac{1}{2}N^2+\frac{1}{2}N \tag{4.19}$$

We therefore say that the back substitution has a computational complexity of (essentially) $N^2/2$. Combining our results, we find that the naive elimination process has a computational complexity of $N^3/3+N^2/2$. Some available methods for solving N linear equations have a complexity of $N!$, so they are of course far less efficient than the Gaussian elimination. (Laboratory participants should compare these complexities for increasing values of N to get a feeling for the difference.)

An interesting assignment (see Exercise 4.15) is the following: After triangularization, it is also possible to find the solution by continuing the permissible operations, this time upward, until the coefficient table has nonzero elements only in its diagonal. This process, called *diagonalization*, is shown in Exercise 4.16 to be less efficient than triangularization followed by back substitution.

We are always, of course, interested not only in a solution algorithm that works but also in its efficiency. This creed will appear repeatedly in laboratory activities, and participants should learn to be critical and be willing to compare various methods of completing a given assignment.

It turns out that when the Gaussian elimination is carried out, various difficulties are encountered because not every system is as naive as has been assumed heretofore. We shall discuss these non-naive cases in the following sections.

4.6 Pivoting

So far we have assumed that at any stage of the triangularization the current element $T(R,R)$ of the operating row (by which we divide), known as the *pivot*, differs from zero and is not even nearly zero. Now we remove this assumption and start with a simple example having the following augmented table:

0	1	1	1	1
1	1	2	1	−1
2	2	4	0	−1
1	2	1	1	2

(4.20)

Because $T(1,1)$ is zero, it cannot be used as a pivot element. A simple way to overcome this difficulty is to interchange the first row and one of the rows below it. This is permissible, of course, because it only reflects the interchange of two of the equations. Interchanging the first and second rows leads to

1	1	2	1	−1
0	1	1	1	1
2	2	4	0	−1
1	2	1	1	2

(4.21)

so now $T(1,1)$ equals 1, and triangularization can be started. The laboratory participants should do just that, noticing that because $T(2,1) = 0$, there is no need to operate on the second row. After operating on the third and fourth rows, we are led to

1	1	2	1	−1
0	1	1	1	1
0	0	0	−2	1
0	0	−2	−1	2

(4.22)

Here again an interchange is needed, this time of the third and fourth rows. This completes the triangularization, and all that remains to be done is the back substitution. Because such interchanges have to be anticipated, the elimination algorithm developed in the previous section must be modified. Before presenting this modification, however, we discuss some other difficulties.

Let us observe that interchanging rows is not always the proper prescription, as seen from the following table (where α is a parameter):

5	5	1	11
2	2	4	α
4	4	7	15

(4.23)

After operating with the first row on the second and third, we obtain

5	5	1.0	11
0	0	3.6	$\alpha - 4.4$
0	0	6.2	6.2

(4.24)

Because $T(2,2)$ and all the elements below it equal zero (here $T(3,2)$ is the only one), no interchange is needed to complete the triangularization. But a difficulty arises when we attempt to carry out the back substitution. From the last row (equation) we immediately get $X(3) = 1$. Then, substituting this $X(3)$ into the second equation, we find ourselves in one of two situations:

(i) $\alpha \neq 8$; say, $\alpha = 7$. In this case, substitution of $X(3) = 1$ into the second equation yields a contradiction. Such a contradiction is reached for every $\alpha \neq 8$, and therefore the system has no solution.

(ii) $\alpha = 8$. In this case, the second equation yields $3.6 = 3.6$. Although this is no contradiction, it does not lead us to $X(2)$. From the first equation we get $X(2) = 2 - X(1)$, thereby using up all the information contained in the system. This system, therefore, has a family of infinitely many solutions; that is, $X(1)$ is arbitrary, $X(2) = 2 - X(1)$, and $X(3) = 1$.

This example illustrates the existence of *singular cases*, in which there is either no solution or there are infinitely many. For those who go on to study linear algebra, this is an excellent point of departure (see Exercises 4.17 and 4.18).

Because cases with vanishing pivots exist, we must modify the elimination algorithm in two aspects. It must provide for the interchange of rows whenever possible [as in (4.20) and (4.21), and similar cases] or, when all the elements below the vanishing pivot are zero, the algorithm should terminate and print an appropriate message.

Having discussed vanishing pivot elements, we now focus our attention on "near singular" cases, in accordance with the principle that "if a numerical procedure actually fails for some values of the data, then the procedure is probably untrustworthy for the values of the data near the failing values". This point is of particular importance when computers are used because, for example, a singular case can easily be mistaken for a nonsingular one, due to round-off errors. To illustrate this point, consider

$$\begin{aligned} X(2) &= 1 \\ X(1) + X(2) &= 2 \end{aligned} \tag{4.25}$$

In this case, after interchanging the rows, we find that $X(1) = X(2) = 1$. The reason for interchanging the rows and not solving directly is that we want to operate in accordance with a general automatic method, which includes a back substitution. A near singular case corresponding to (4.25) is

$$\begin{aligned} \varepsilon X(1) + X(2) &= 1 \\ X(1) + X(2) &= 2 \end{aligned} \tag{4.26}$$

where ε is a small positive number. Triangularizing (4.26), we find

$$\begin{aligned} \varepsilon X(1) \quad + \quad X(2) &= 1 \\ \left(1 - \frac{1}{\varepsilon}\right) X(2) &= 2 - \frac{1}{\varepsilon} \end{aligned} \tag{4.27}$$

Now, solving backward, as the computer would do, we reach

$$\begin{aligned} X(2) &= \frac{2 - \dfrac{1}{\varepsilon}}{1 - \dfrac{1}{\varepsilon}} \\ X(1) &= \frac{1 - X(2)}{\varepsilon} \end{aligned} \tag{4.28}$$

For sufficiently small ε, the number $1/\varepsilon$ is so large that $(1-1/\varepsilon)$ and $(2-1/\varepsilon)$ are indistinguishable as far as the computer is concerned. Consider, for example, a microcomputer with eight decimal digits. For $\varepsilon = 10^{-9}$, say, the result of computing $2 - 1/\varepsilon$ rounded to eight decimal digits is -10^9. Similarly, $1 - 1/\varepsilon$ is found to equal -10^9, and consequently, the computed solution is $X(2) = 1$ and $X(1) = 0$. Substitution of these values into (4.26) shows that the second equation is not satisfied even approximately. If we multiply both numerator and denominator of the first equation of (4.28) by ε and substitute the result into the second, we obtain

$$\begin{aligned} X(2) &= \frac{1-2\varepsilon}{1-\varepsilon} \approx 1 \qquad \text{(to eight decimal digits)} \\ X(1) &= \frac{1}{1-\varepsilon} \approx 1 \qquad \text{(to eight decimal digits)} \end{aligned} \tag{4.29}$$

which satisfy (4.26) to a high order of accuracy. However, as we have seen, the computation of the solution by means of (4.28) – as the computer would have done it by means of naive elimination – leads to a serious error. This error stems from division by the small number ε (in the course of carrying out the naive elimination), which creates a very large number $1/\varepsilon$ that eventually generates the erroneous solution. The interchange of rows, used for zero pivot elements, turns out to be the correct remedy even for pivot elements that are only nearly zero, as in (4.26). It is therefore advisable always to interchange rows, so as to obtain the largest (in absolute value) possible pivot element. Thus, interchanging rows in (4.26), we obtain

$$\begin{aligned} X(1) + X(2) &= 2 \\ \varepsilon X(1) + X(2) &= 1 \end{aligned} \tag{4.30}$$

Triangularizing and solving backward, we now obtain the satisfactory solution

$$\begin{aligned} X(2) &= \frac{1-2\varepsilon}{1-\varepsilon} \approx 1 \\ X(1) &= 2 - X(2) \approx 1 \end{aligned} \tag{4.31}$$

Collecting our results, we now modify the elimination algorithm. We add a number of instructions to interchange rows so that in each cycle of the operating-row loop the largest (in absolute value) pivot element is used. This technique is referred to as *pivoting*. The modification must also detect singular cases in which the pivot element and all the elements below it are zero, print an appropriate message, and terminate. We suggest constructing a special procedure for this purpose, to be invoked between Steps 9 and 10 in the triangularization algorithm. This *pivoting procedure* can take the following form:

1. Set PIVO = $|T(R,R)|$ and $M = R$.
2. Repeat Steps 2–4 for L increasing from $R+1$ to N.
3. If $|T(L,R)| >$ PIVO, set PIVO = $|T(L,R)|$ and $M = L$.

4. Close the loop started in Step 2.
5. If PIVO = 0, print an appropriate message and terminate.
6. If $M = R$, go to Step 12.
7. Repeat Steps 7–11 for J increasing from R to $N + 1$.
8. Set HOLD = $T(R, J)$.
9. Replace the value of $T(R, J)$ by $T(M, J)$.
10. Replace the value of $T(M, J)$ by HOLD.
11. Close the loop started in Step 7.
12. End.

As we shall see, computational difficulties may arise whenever |PIVO| is very small. The "smallness" crucially depends on the intrinsic accuracy of the computer being used. To demonstrate this, let us consider the following system:

−1.41	2	0	1
1	−1.41	1	1
0	2	−1.41	1

(4.32)

whose solution (correct to three significant figures) is $X(1) = X(2) = X(3) = 1.69$. Now let us make the artificial assumption that we are equipped with a three-digit computer. Thus, to simulate our assumed computing device, we round off every computed result to three significant figures. When we start the triangularization, we multiply the first row by $1/1.41 = 0.709$ and add it to the second. Then we multiply the new second row by $-2/0.01 = -200$ and add it to the third, to get

−1.41	2	0	1
0	0.01	1	1.71
0	0	−201	−341

(4.33)

The back substitution now yields $X(3) = 1.70$, $X(2) = 1.00$, and $X(1) = 0.709$, which is manifestly incorrect. Had we used the pivoting strategy, the correct solution (to three significant figures) would have been found (see Exercise 4.19). Pivoting would not have been necessary, though, had we used a five-digit computer. With such a computer, we multiply the first row by $1/1.41 = 0.70922$ and add it to the second, then multiply the second by $-2/0.00844 = -236.97$ and add it to the third, to get

$$\left[\begin{array}{ccc|c} -1.41 & 2 & 0 & 1 \\ 0 & 0.00844 & 1 & 1.7092 \\ 0 & 0 & -238.38 & -404.03 \end{array}\right] \tag{4.34}$$

This time the back substitution gives $X(3) = 1.6949$, $X(2) = 1.6943$, and $X(1) = 1.6941$, which is clearly a vast improvement. Moreover, repeating the process with eight-digit accuracy, we obtain $X(3) = 1.6949153$, $X(2) = 1.6949262$, and $X(1) = 1.6949307$, which is even closer to the true solution given by $X(1) = X(2) = X(3) = 1/0.59$.

These results suggest the following conclusion: The degree of computational difficulties depends on the intrinsic accuracy of the computer being used. Pivoting helps overcome those difficulties (except for situations to be described). Moreover, higher intrinsic precision also helps (many computers enable the user to switch from single to double precision). Further ideas on the pivoting concept and its implementation as part of Gaussian elimination can be found in Exercises 4.20–4.22.

4.7 Scaling

As noted at the end of the preceding section, pivoting is not always the remedy for computational difficulties. Consider, for example,

$$\begin{aligned} 10^{-6}X(1) \quad - \quad X(2) &= 1 \\ 10^{-6}X(1) + 10^{-12}X(2) &= 10^{-6} \end{aligned} \tag{4.35}$$

The modified elimination algorithm will not find it necessary to interchange the rows, and thus the triangularization gives

$$\begin{aligned} 10^{-6}X(1) \quad - \quad X(2) &= \ \ 1 \\ (1 + 10^{-12})X(2) &= -1 + 10^{-6} \end{aligned} \tag{4.36}$$

Suppose now that the computations are performed with a six-digit computer. The backward substitution then gives $X(2) = -1$ and $X(1) = 0$. In a six-digit computer the numbers 10^{-6} and 10^{-12} are so small that when added to 1, say, they are negligible. In fact, for such a computer, both $1 + 10^{-6}$ and $1 + 10^{-12}$ equal 1, as would have been the case for weighing an elephant and reweighing it with a fly on its back. The laboratory participants should realize, however, that they cannot neglect 10^{-6} and 10^{-12} in (4.35). In a six-digit floating-point computation, such numbers are negligible only when added to, or subtracted from, numbers larger then themselves by six or more orders of magnitude. If we compute the solution of (4.35) to a high accuracy, we find that $X(1) = 1 + 10^{-6}$ and $X(2) = -1 + 10^{-6}$. This solution rounded off to six significant digits is $X(1) = 1$ and $X(2) = -1$, which shows how erroneous the previously computed solution was. This phenomenon is associated with the fact that the coefficients of the given system are "imbalanced." If we examine the absolute values of the coefficients, we find that the maximal value in the first row is 1 whereas in the second row it is 10^{-6}. If we multiply the second row by 10^6, before we start the solution process, we will have a system whose maximal coefficient (in absolute value) in each row is 1, that is, a balanced system. Such an introductory procedure will be referred to as *scaling*. After scaling (4.35), we get

$$\begin{aligned} 10^{-6}X(1) \quad - \quad X(2) &= 1 \\ X(1) + 10^{-6}X(2) &= 1 \end{aligned} \tag{4.37}$$

Now the modified elimination algorithm will interchange the two equations, triangularize the system, and yield

$$\begin{aligned} X(1) \quad + \quad 10^{-6}X(2) &= 1 \\ (-1 - 10^{-12})X(2) &= 1 - 10^{-6} \end{aligned} \tag{4.38}$$

Back substitution (with six-digit accuracy) leads to $X(2) = -1$ and $X(1) = 1$, which is the true solution to this order of accuracy. Therefore, scaling should be added to the modified elimination algorithm, as a separate procedure, to be invoked (when needed) before triangularization takes place. The laboratory participants should experiment with scaling and see for themselves its effect on pivoting, that is, on the *row interchange strategy*. In this connection see Exercise 4.23.

4.8 Ill-conditioning

When we attempt to solve a given linear system on a computer, round-off errors are always present in the initial data or in intermediate results. In addition, the data may contain errors stemming from the use of measurements with limited accuracy. It is necessary to investigate the influence of such errors on the computed solution. In particular, we want to watch for systems in which small errors in the data will cause large changes in the computed solution. Such systems are said to suffer from *ill-conditioning.* Consider, for example, the system (suggested by Jerome Dancis) whose triangular augmented table is the following:

$$\begin{array}{|ccccc|c|} 0.1 & -1 & 0 & 0 & 0 & 0.1 \\ 0 & 0.1 & -1 & 0 & 0 & -1 \\ 0 & 0 & 0.1 & -1 & 0 & 0.1 \\ 0 & 0 & 0 & 0.1 & -1 & -1 \\ 0 & 0 & 0 & 0 & 0.1 & 0.1 \end{array} \tag{4.39}$$

Using back substitution, we obtain $X(5) = 1$, $X(4) = 0$, $X(3) = 1$, $X(2) = 0$, and $X(1) = 1$. If we change the rightmost column a little, so that $T(5,6) = 0.101$ and all the other coefficients remain unchanged, the solution will be $X(5) = 1.01$, $X(4) = 0.1$, $X(3) = 2$, $X(2) = 10$, and $X(1) = 101$. Thus, a change of 1 percent in one coefficient causes a considerable change in the solution. This system has an obvious pattern; thus, the laboratory participants can construct larger, similar systems in which this phenomenon is even more pronounced (see Exercise 4.24). Such systems are clearly ill-conditioned.

Next we illustrate the effect of rounding off the initial data that is being stored in the computer's memory. To this end, we consider the following augmented table:

$$\begin{array}{|ccc|c|} 1 & \frac{1}{2} & \frac{1}{3} & 1 \\ \frac{1}{2} & \frac{1}{3} & \frac{1}{4} & 0 \\ \frac{1}{3} & \frac{1}{4} & \frac{1}{5} & 0 \end{array} \tag{4.40}$$

The solution of this system is $X(1) = 9$, $X(2) = -36$, and $X(3) = 30$. Let us now simulate a two-digit computer by rounding off every intermediate result to two significant figures. This simulation enables us to illustrate this kind of ill-conditioning using just three equations. Because the pattern of this example is obvious, the laboratory participants can easily construct a similar larger system, solve it using the full accuracy of their computing device, and observe the rounding-off effect (see Exercise 4.25). In two-digit accuracy, (4.40) becomes

$$\left[\begin{array}{ccc|c} 1.00 & 0.50 & 0.33 & 1 \\ 0.50 & 0.33 & 0.25 & 0 \\ 0.33 & 0.25 & 0.20 & 0 \end{array}\right] \qquad (4.41)$$

We have caused an error of about $\frac{3}{1000}$ in $T(1,3)$, $T(2,2)$, and $T(3,1)$, and the solution obtained is $X(1) = -1.1$, $X(2) = 20$, and $X(3) = -24$, differing considerably from the true solution. Had we used five-digit computations, we would have obtained a solution very close to the true one. The very fact that an increase in the accuracy of the computations changes the solution by so much implies that our system suffers from ill-conditioning. In practice, however, this indication of ill-conditioning comes a posteriori, that is, when a solution previously computed is recomputed with greater precision. What we would like, though, is an earlier indication of ill-conditioning, pointing to the source of the trouble. We shall come back to this issue in the next section.

To gain further insight into ill-conditioning, consider the system

$$\begin{aligned} 2X(1) + 6.000000X(2) &= 8.000000 \\ 2X(1) + 6.000001X(2) &= 8.000001 \end{aligned} \qquad (4.42)$$

whose unique solution is $X(1) = X(2) = 1$. Changing $T(2,2)$ from 6.000001 to 5.999999, but leaving all other data unchanged, leads to the solution $X(2) = -1$ and $X(1) = 7$. Examination of (4.42) reveals that both equations contain almost the same information and are represented graphically by two straight lines that are almost coincident. Thus, a very slight change in the lines shifts their intersection point considerably, as can be seen in Figs. 4.1 and 4.2, where $x = X(1)$ and $y = X(2)$.

We want to emphasize the distinction between errors stemming from the nature of a system (ill-conditioning) and errors resulting from the method of solution. The latter can be treated by modifying the algorithm (scaling and pivoting) whereas ill-conditioning is intrinsic and cannot be removed even by a sophisticated modification. In an ill-conditioned case, the solution is hypersensitive to changes in the data. Furthermore, if we substitute the computed solution in the given equations, we will find that even though the equations are satisfied to a high accuracy, the computed solution column many differ considerably from the true solution column.

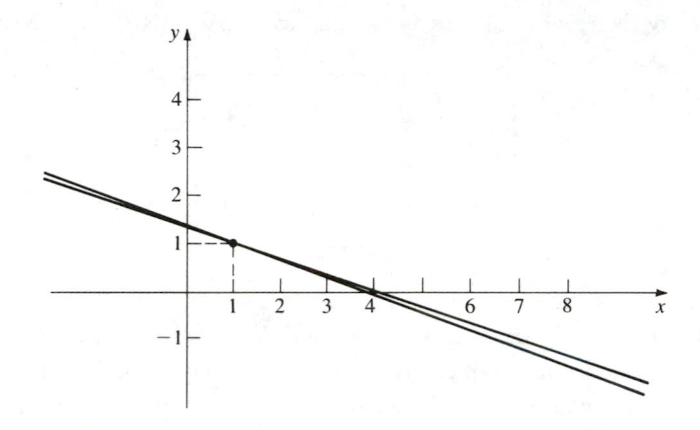

Fig. 4.1

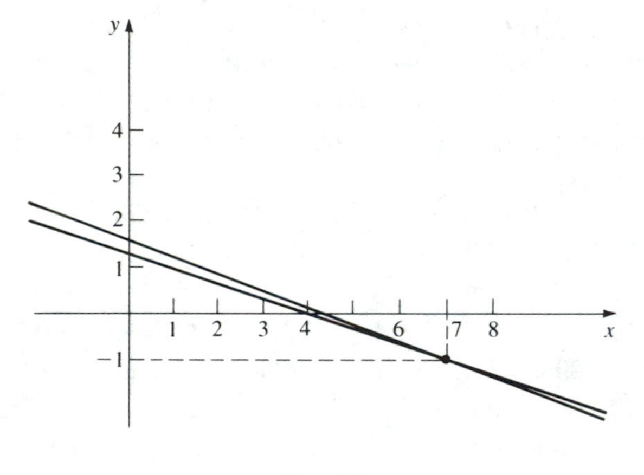

Fig. 4.2

To see this more clearly, let us define the *residual column* R as the difference between the left-hand side of the system, after substituting the computed solution column in it, and the right-hand side column. We also define the *error column* E as the difference between the computed solution column and the true (exact) solution column. A system is considered ill-conditioned if some components of E are very large compared with the largest (in absolute value) component of R. In this connection see Exercise 4.26.

At this point, the laboratory instructor can raise the question of what is meant by a good computed solution. The following ill-conditioned example by George Forsythe may be helpful:

$$\begin{aligned} 0.780X(1) + 0.563X(2) &= 0.217 \\ 0.913X(1) + 0.659X(2) &= 0.254 \end{aligned} \tag{4.43}$$

The true (exact) solution of this system is $X(1) = 1$ and $X(2) = -1$. Suppose we examine the pair of numbers 0.999 and -1.001 as a possible solution. This solution is very close to the true solution, and its residual column consists of $R(1) = -1.343 \cdot 10^{-3}$ and $R(2) = -1.572 \cdot 10^{-3}$. On the other hand, the components of the residual column generated by the pair 0.341 and -0.087, which does not even come close to the true solution, are $R(1) = -10^{-6}$ and $R(2) = 0$. This is so, even though the components of the error column E are much smaller in the first case than in the second. Usually, we are interested in small components of E, but sometimes small components of R have first priority, depending on the application involved. Thus, a "good computed solution" can be meaningfully characterized only in terms of the intended application. On the other hand, for systems that are far from being ill-conditioned (well conditioned, so to speak), no such ambiguity arises because the components of R and those of E are very small. In this connection see Exercise 4.27.

4.9 Concluding remarks

Let us summarize the various situations with which we may be confronted when we attempt to solve linear systems using the full algorithm developed in this chapter.

1. In the *naive case*, the solution process is carried out without any difficulties, computational or otherwise.
2. If the *elements of the coefficient table are imbalanced*, scaling should be applied.
3. If a *pivot element is small (in absolute value) relative to the elements below it in the same column*, interchanging of rows is needed; that is, the pivoting procedure should be invoked.
4. In the *singular case*, in which a pivot element and all the elements below it in the same column are zero, so that no interchange of rows is useful, the algorithm should print out an appropriate message and terminate.
5. In *ill-conditioned systems*, the ill-conditioning cannot be removed by

row-interchange, scaling, or otherwise; but double-precision computations can be helpful and can serve as an a posteriori indication. An early indication of ill-conditioning occurs if at some stage of the triangularization *the pivot element and all elements below it are small.* The pivoting procedure should detect such a situation (see Exercise 4.28) and print a warning message.

The triangularization is no doubt the main part in the algorithm we have developed, consuming most of the computing time. With a view toward future development of computing, we should examine the possibility of carrying out various parts of the algorithm in parallel. In other words, suppose a computer can be thought of as a collection of small subcomputers that operate simultaneously. For such computers, it is interesting to investigate whether the elimination algorithm can be changed so as to take advantage of the parallel computing and thus increase efficiency. Indeed, when a row operates on a row below it, it can operate on the other rows below it in parallel. In addition, when a row operates on another row, the operations on its various elements can also be carried out in parallel. Because parallel computing has the potential of cutting computing time tremendously and becoming predominant in modern numerical mathematics, laboratory participants should keep this idea in the back of their minds throughout their work.

Exercises

4.1 Perform the triangularization steps on the pharmacist problem, represented by (4.3). Start by multiplying the first row of (4.3) by (−4) and adding it to the second. Continue analogously until triangularization is completed.

4.2 Repeat Exercise 4.1 for the system whose augmented table is

4	2	1	1	−3
8	12	3	4	6
2	1	4	1	6
−4	−2	−1	6	10

4.3 Find the solution of the pharmacist problem by performing back substitution on the triangular form you obtained in Exercise 4.1.

4.4 Find the solution of the system in Exercise 4.2 by back substitution, making use of the triangular form you obtained there.

4.5 The triangularizations performed in the text and in Exercises 4.1 and 4.2 are known as upper triangularizations. Reconsider the pharmacist problem (4.3), and carry out a lower triangularization by first multiplying the last (third) row by (-2), and adding it to the row above it (the second), and so on. When this is done, solve the system by forward substitution.

4.6 Repeat Exercise 4.5 for the 4×4 system in Exercise 4.2.

4.7 Explain carefully why it is sufficient (and more efficient) to let the index C of the innermost loop (of the triangularization) start at R rather than at 1.

4.8 Write a program for the solution of a naive $N \times N$ linear system, and run it so as to solve the pharmacist problem (4.2) and the 4×4 system given in Exercises 4.2 and 4.4.

4.9 Run the Gaussian elimination program of Exercise 4.8 so as to solve the 5×5 system for which $T(R,C) = C^{R-1}$, $1 \leq R \leq 5$ and $1 \leq C \leq 6$.

4.10 Repeat Exercise 4.9, for an analogous 6×6 system.

4.11 Run the Gaussian elimination program for the system

30	6	3	1	90
1	12	1	2	72
2	5	57	3	57
4	3	6	43	86

Once you have the solution, compare the rightmost column with the diagonal elements $T(1,1)$, $T(2,2)$, $T(3,3)$, $T(4,4)$. What can one say about the solution of this special system?

4.12 Write a program that solves a general $N \times N$ linear system by lower triangularization followed by forward substitution, as indicated in Exercise 4.5.

4.13 Run the program you wrote in Exercise 4.12 for the pharmacist problem and the 4×4 system of Exercise 4.2. Compare the results with the hand-calculated results you obtained in Exercises 4.5 and 4.6.

4.14 Show that the computational complexity of triangularization is essentially $N^3/3$, rather than $N^3/2$ as claimed by (4.18). To derive this result, you will have to use the formulas for the sum of successive integers and the sum of their squares.

4.15 Write a diagonalization algorithm in the following way: Start with the upper triangularization algorithm, then follow with permissible operations that actually perform a lower triangularization, so that all nondiagonal elements of the coefficient table become zero. Write a corresponding program, and run it to solve the system given in Exercise 4.11.

4.16 Explain why the diagonalization algorithm is less efficient than triangularization followed by back substitution.

4.17 Show that the system whose augmented table is

4	2	1	1	3
24	12	3	4	20
8	4	18	16	−8
4	2	1	−6	10

has a family of infinitely many solutions, in which the value of one of two unknowns (which one?) can be chosen arbitrarily.

4.18 Show that the system whose augmented table is

1	1	1	1	10
2	−1	3	−4	−7
4	1	5	−2	13
−1	2	−2	5	17

has a family of infinitely many solutions in which any two unknowns can be chosen arbitrarily. Moreover, show that if, for example, the number 17 in the rightmost column is changed, there is no solution.

4.19 Reconsider the system (4.32) whose solution, correct to three digits, is $X(1) = X(2) = X(3) = 1.69$. Solve this system by hand, using Gaussian elimination with pivoting, and maintaining three-digit accuracy throughout. Compare your results to the correct solution just cited, as well as to the incorrect one – obtained without pivoting – following (4.33) in the text.

4.20 Write a pivoting procedure, to be invoked at the appropriate place, in a program that solves a linear system by lower triangularization followed by forward substitution, as in Exercise 4.12.

4.21 Reconsider the system given in Exercise 4.11, and solve it using the modified Gaussian elimination algorithm that contains a pivoting procedure. Will the program actually invoke the pivoting procedure in this case? Explain why! (Such a system is known as *diagonally dominant*.)

4.22 Write your own 3×3 diagonally dominant system. Interchange the first and third rows, then solve the new system using Gaussian elimination with pivoting. Construct an "original" triangularization that will solve this new system without invoking pivoting at all.

4.23 Consider the system whose augmented table is given by

0.1	−1	1	−1
20	40	2000	1840
8	15	200	135

Triangularize this system by hand, using Gaussian elimination with pivoting, and record the stages at which pivoting has taken place. Next, reconsider the given system, this time with scaling (the first row, of course, needs no scaling). Triangularize the system and observe how scaling caused a different pivoting strategy.

4.24 Construct a 7×7 system that is similar to (4.39) in the text, and find its solution. Next change $T(7,8)$ from 0.1 to 0.101, leaving all other entries unchanged, and observe the dramatic change in the solution.

4.25 Construct a 9×9 system that is similar to (4.40) in the text [i.e., $T(R,C) = 1/(R+C-1)$ in the coefficient table]. Solve this system using single precision, then solve it again using double precision, and compare the results.

4.26 Consider the system whose augmented table is given by (4.39) in the text, and assume that, due to round-off errors, the solution was computed to be $X(1) = 101$, $X(2) = 10$, $X(3) = 2$, $X(4) = 0.1$, and $X(5) = 1.01$. Compute the residual column R and – remembering that for the exact solution we have $X(1) = X(3) = X(5) = 1$ and $X(2) = X(4) = 0$ – the error column E

as well. Finally, find the ratio of the maximal component of E to the maximal component of R. Note that this ratio is huge, which is characteristic of ill-conditioning. Repeat the entire calculation for the 7×7 system you constructed in Exercise 4.24.

4.27 Find the components of the residual column R and the error column E, after you have numerically solved the system whose augmented table is given by

4.4	2.1	2.5	8.0
2.8	3.9	2.3	8.0
2.1	2.7	4.2	8.0

The exact solution of this system is $X(1) = X(2) = X(3) = \frac{8}{9}$. Note that this system does not suffer from the ill-conditioning observed in Exercise 4.26.

4.28 Suppose you are solving a linear system, using a Gaussian elimination algorithm with scaling and pivoting. How should the pivoting procedure be modified in order to detect ill-conditioning and issue a timely warning?

5

Algorithmic computations of π and e

5.1 Introduction

As our next numerical laboratory assignment, we consider a variety of methods for the computation of π and e. The number π plays a central role throughout mathematics, and its actual computation to any desired accuracy should be part of the mathematical foundations acquired by a college graduate. Practical experience shows that this subject arouses a lot of interest among people who are familiar with the number π and curious about how to obtain its value to a desired accuracy. The number π enjoys a special status in the minds of the students, who get a kick out of unfolding its mystery. Later, we examine the genesis of the number e, as well as its computation.

The four methods we present for the computation of π cover a range of mathematical concepts. We stress that the mere computation of π is by no means the only feature of these methods, though that may appear to be so at first sight. Rather, we shall see that each method generates beneficial by-products. Analysis of the effects of round-off errors, rates of convergence, probabilistic reasoning, computer simulation, and so forth will surface in a natural way and give the laboratory participants a good dose of mathematics while their attention is focused on the computation of π.

5.2 The method of Archimedes

We first calculate π using a variant of the *method of Archimedes*, based on polygons inscribed in and circumscribed about the unit circle. It is assumed that π is known to the laboratory participants, from geometry, as the constant ratio of the circle's circumference to its diameter.

Let us start with a unit circle, about which a regular hexagons is circumscribed and in which a corresponding regular hexagon is inscribed; each hexagon can be divided into six equilateral triangles. The side a_6 of the inscribed hexagon equals unity because it equals the radius of the circle. The side c_6 of the circumscribing hexagon is equal to $\frac{2}{3}\sqrt{3}$ by the theorem of Pythagoras (see Exercise 5.1). Accordingly, comparing half the circle's perimeter with the perimeters of the hexagons, we obtain an initial estimate of π:

$$3 = 3a_6 < \pi < 3c_6 = 2\sqrt{3} < 3.47 \tag{5.1}$$

Clearly, we can now double the number of sides of our hexagons (giving a 12-sided polygon) and repeat the process. In this way we eventually arrive at an inscribed $(2n)$-gon with side a_{2n}, whose value will be shown to be given recursively in terms of a_n by

$$a_{2n} = \sqrt{2 - \sqrt{4 - a_n^2}} \qquad a_6 = 1 \tag{5.2}$$

Equation (5.2) is obtained from the application of the Pythagorean theorem to the triangles with sides $(a_n/2,\, b_n,\, 1)$, and $(a_n/2,\, 1 - b_n,\, a_{2n})$, respectively, in Fig. 5.1. Thus,

$$b_n = \sqrt{1 - \frac{a_n^2}{4}} \tag{5.3}$$

and

$$a_{2n} = \sqrt{\frac{a_n^2}{4} + (1 - b_n)^2} \tag{5.4}$$

which leads to (5.2). Equation (5.2) is the key relation, enabling us to "climb" from n to $2n$ and obtain the semiperimeter na_{2n} of the $(2n)$-gon in terms of its predecessor $(n/2)a_n$.

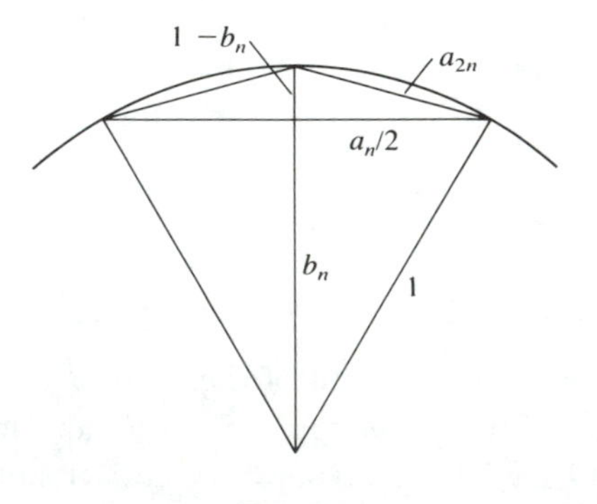

Fig. 5.1

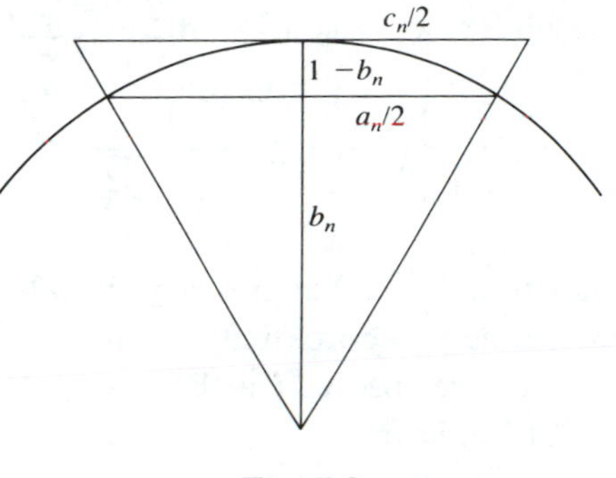

Fig. 5.2

Next, consider the circumscribing n-gons with sides c_n. From Fig. 5.2 and the similarity of triangles, we obtain (see Exercise 5.2)

$$\frac{\frac{c_n}{2}}{1} = \frac{a_n/2}{b_n} \tag{5.5}$$

which, in conjunction with (5.3), yields

$$c_n = \frac{a_n}{\sqrt{1 - \frac{a_n^2}{4}}} \tag{5.6}$$

Collecting our results, we now relate the semiperimeters through

$$\frac{n}{2}a_n \leq \pi \leq \frac{n}{2}c_n = \frac{\frac{n}{2}a_n}{\sqrt{1 - \frac{a_n^2}{4}}} \tag{5.7}$$

It is evident from (5.7) that the number π is sandwiched between the values of the semiperimeters $(n/2)a_n$ and $(n/2)c_n$. Moreover, the sequence $3a_6, 6a_{12}, 12a_{24}, \ldots$ of the semiperimeters of the inscribed polygons is monotonically increasing (and bounded) whereas the sequence $3c_6, 6c_{12}, 12c_{24}, \ldots$ of the semiperimeters of the circumscribed polygons is monotonically decreasing (and bounded). These facts are based on elementary geometry (see Exercise 5.3). It is suggested that the laboratory participants compute the extreme sides of (5.7) for $n = 6$, 12, 24, 48 to obtain numerical corroboration for their intuitive feeling and geometric reasoning.

To estimate the rate at which the extreme sides of (5.7) zero-in on π, let us subtract $(n/2)a_n$ from each of the three expressions in (5.7), and then divide each resulting expression by $(n/2)a_n$. This yields

$$0 \leq \frac{\pi - \left(\frac{n}{2}\right) a_n}{\left(\frac{n}{2}\right) a_n} \leq \frac{1}{\sqrt{1 - \frac{a_n^2}{4}}} - 1 \tag{5.8}$$

Note that the middle term in (5.8) is precisely the relative error incurred when $(n/2)a_n$ is taken as an approximation for π. To estimate the extreme right side of (5.8), we observe (see Exercise 5.4) that for values of u satisfying $0 \leq u \leq 1$, we have

$$\frac{1}{\sqrt{1 - \frac{u}{4}}} - 1 \leq \frac{u}{6} \tag{5.9}$$

As we saw earlier, $a_6 = 1$, and clearly a_n is decreasing with n. Consequently, $0 \leq a_n^2 \leq 1$, so we identify u in (5.9) with a_n^2 and rewrite the estimate (5.8) as

$$0 \leq \frac{\pi - \left(\frac{n}{2}\right) a_n}{\left(\frac{n}{2}\right) a_n} \leq \frac{a_n^2}{6} \tag{5.10}$$

Now it is clear that the perimeter na_n of the inscribed polygon is smaller than the perimeter nc_n of the circumscribed polygon. The latter is decreasing with n, and in conjunction with (5.1) we thus have

$$na_n \leq nc_n \leq 4\sqrt{3} \tag{5.11}$$

From (5.11) and (5.10), we obtain our final estimate of the relative error in the form

$$0 \leq \frac{\pi - \left(\frac{n}{2}\right) a_n}{\left(\frac{n}{2}\right) a_n} \leq \frac{8}{n^2} \tag{5.12}$$

The error decay is thus seen to be inversely proportional to n^2, that is, to the square of the number of sides of the polygons. Indeed, because here $n = 6 \cdot 2^k$ for $k = 0, 1, 2 \ldots$ starting with a hexagon, the error bound in (5.12) equals $8/(36 \cdot 2^{2k})$. Thus, to obtain q correct figures, we require

$$\frac{8}{36 \cdot 2^{2k}} \leq \frac{1}{2} \cdot 10^{-q} \tag{5.13}$$

or

$$\frac{2}{3} \cdot 10^{q/2} \leq 2^k \tag{5.14}$$

Here k is the number of Archimedes cycles, in each of which the number of sides of the polygons is doubled. Now because $10 \leq 2^{10/3}$, it follows that $k \geq \frac{5}{3}q$ is more than enough to guarantee q correct figures.

Before we write the preceding ideas in algorithmic form, we want to say that it is by no means essential to start with hexagons; squares, for example, would have been equally suitable, requiring minor modifications of the error estimates (see Exercise 5.5). In all cases, the upper and lower Archimedean estimates of π approach each other – zeroing-in on π – at a rate inversely proportional to n^2 (i.e., to 2^{2k}).

The preceding results can now be cast in algorithmic form, programmed, and executed on the laboratory microcomputers. The following algorithm is suggested:

1. Input the value of q (desired number of correct figures).
2. Set $t = \frac{1}{2} \cdot 10^{-q}$, $a = 1$, and $n = 6$.
3. Replace the value of n by $2n$.
4. Replace the value of a by $\sqrt{2 - \sqrt{4 - a^2}}$.
5. Compute $L = n \cdot a/2$ and $U = L/\sqrt{1 - a^2/4}$.
6. Compute $P = (U + L)/2$ and $E = (U - L)/2$.
7. Print the values of n, L, U, E, and P.
8. If $E \geq t$, return to Step 3.
9. End.

The translation of this algorithm into the programming language used in the laboratory should be accompanied by the appropriate discussion of its various steps. We initialized $n = 6$ because we started with hexagons. The quantities L and U in Step 5 constitute the lower and upper approximations to π, in line with (5.7). The quantity P is the average of L and U, and hence its deviation from π will never exceed E. Step 7 prints the current values of the relevant quantities and exhibits the gradual improvements as the upper and lower approximations zero-in on π. Thus, more and more correct decimal figures of π are generated, as can be vividly seen by the laboratory participants. Finally, the process will stop by means of Step 8, when the error bound E is within the prescribed tolerance t. We could have used a different strategy for the termination of the loop, by using our a priori results obtained earlier and letting n increase from 6 to $6 \cdot 2^k$, where the total number k of cycles is the first integer not less than $\frac{5}{3}q$.

We ran this algorithm on a computer with $q = 5$ and obtained Table 5.1, in which we display seven-decimal figures. From the table we find that (correct to five decimals) $\pi = 3.14159$. As mentioned, we could have started the process of approximating π with circumscribed and

Table 5.1. *Archimedean approximations of* π

n	L	U	E	P
12	3.1058286	3.2153903	0.1095617	3.1606094
24	3.1326286	3.1596599	0.0270329	3.1461443
48	3.1393502	3.1460862	0.0067360	3.1427182
96	3.1410319	3.1427146	0.0016828	3.1418733
192	3.1414525	3.1418731	0.0004206	3.1416628
384	3.1415576	3.1416627	0.0001051	3.1416102
768	3.1415839	3.1416102	0.0000263	3.1415970
1536	3.1415905	3.1415970	0.0000066	3.1415937
3072	3.1415921	3.1415937	0.0000016	3.1415930

circumscribing squares. The algorithm would then have to be slightly modified, as indicated in Exercise 5.6.

We re-ran the preceding algorithm on a large computer (single precision, 15 decimal digits) with $q = 8$ and obtained $\pi = 3.14159265$. However, when we wanted to obtain even greater accuracy by increasing q, we began to encounter a loss of accuracy. For computers with smaller intrinsic accuracy, such "contamination" would set in at an earlier stage. This loss of accuracy is not merely an interesting subject in itself, it can also serve as a point of departure for the study of round-off errors and their effect and importance in actual computations. There is no better place for the study of this phenomenon than the mathematical laboratory, and we shall turn our attention to this subject in the next section.

5.3 Round-off errors and loss of accuracy

The loss of accuracy we encountered at the end of Section 5.2 when we increased q to obtain greater accuracy, manifested itself as false additional figures for the approximation of π. The fact that these additional figures were false and had therefore contaminated the expected closer approximation of π was confirmed when the results were compared with a very accurate value of π obtained by other methods (to be discussed later). To see the onset and propagation of this loss of accuracy, let us replace the condition $E \geq t$ in Step 8 of the algorithm by the condition $n < M$ for some suitably large M. In doing so, we are actually using polygons with an ever-increasing number of sides, but for the time being, we disregard the issue of accuracy. We ran this modified algorithm for

Table 5.2. *Contaminated approximations*

n	P	
12	3.1	60609434
24	3.14	6144283
48	3.14	2718216
96	3.141	873285
192	3.141	662761
384	3.141	610182
768	3.14159	7033
1536	3.14159	3742
3072	3.141592	957
6144	3.141592	723
12288	3.1415926	14
24576	3.14159265	3
49152	3.141592	318
98304	3.141592	402
196608	3.1415	86844
393216	3.141	499418
786432	3.14	0924553

$M = 800,000$ on a large computer with an intrinsic accuracy of fifteen decimal digits and obtained Table 5.2 of n and P displayed to nine decimal figures. We left extra blank space between the last accurate figure of π (obtained later by better methods) and the first inaccurate one. The table shows how the number of accurate figures first increases with increasing n and then decreases, as contamination due to accumulation of round-off errors sets in, as will be explained. It goes without saying that the Archimedean method cannot be blamed for the loss of accuracy. Rather, the implementation of the method on the computer is at fault. Indeed, when we ran the algorithm on a PC with nine decimal figures, contamination set in much earlier, as seen in Table 5.3. The sensitivity of the results to the internal accuracy of the computing device shows that the variant of the Archimedean algorithm presented here is *numerically unstable*. At this point, the laboratory participants should write and run their own modified program (for suitably large M) and observe the contamination as it develops on their microcomputers (see Exercise 5.7).

Round-off errors are naturally present in any computing device because it carries only a finite number of digits. In some computations, the compounded effects of such errors are limited to the last figures and are therefore of no major concern. However, in cases typified by our

Table 5.3. *Early onset of contamination*

n	P	
12	3.1	60609
24	3.14	6144
48	3.14	2718
96	3.141	873
192	3.141	663
384	3.1415	98
768	3.14159	3
1536	3.1415	83
3072	3.141	487
6144	3.141	862
12288	3.14	4864
24576	3.1	56844

Archimedean algorithm, small errors build up into large errors and gradually destroy the accuracy of the results, in extreme situations rendering them totally meaningless.

If, as in Chapter 1, we are given $A = 0.4564988$ and $B = 0.4564976$, in which the last (seventh) figure of each number may be inaccurate (say, due to round-off errors), and if we then compute $C = A - B$, we obtain 0.0000012, in which the second significant figure is already in error. Subtracting one number from another slightly different number has generated a considerable loss of accuracy. Our Archimedean algorithm suffers from just this kind of malady. The expression $(2 - \sqrt{4 - a^2})$ in Step 4, which plays an essential role in the algorithm, contains the value of a (the side of the polygon), which is very small indeed for large n. Thus, $\sqrt{4 - a^2}$ differs very slightly from 2, and when we subtract it from 2, we encounter the phenomenon just described. The upshot is that instead of obtaining accurate results by using polygons with many thousands of sides, we came up with the same figure, 3.14, already obtained by Archimedes by drawing polygons on the beach.

Laboratory participants should be made aware that small errors accompany most computations, whether they are round-off errors or errors due to data supplied by inaccurate measurements. Thus, it is vital to recognize and beware of numerical instabilities. If possible, we should look for numerically stable algorithms. As a matter of fact, there is another variant of the Archimedean method based on inscribed and circumscribed regular polygons, in which the relevant formulas are derived by using the properties of similar triangles (see Arthur Engel, *Elementary*

Mathematics from an Algorithmic Standpoint, p. 71). The algorithm based on this variant is numerically stable because it does not contain problematic subtractions of the kind just described. We presented the unstable variant because we want laboratory participants to become familiar with the computational phenomenon it represents. In the next section we shall take up a different method for the computation of π, based on area approximation, in the spirit of the ideas and laboratory assignments of Chapter 3.

5.4 Approximations through areas

We shall now approximate π in the mathematical laboratory by finding the related area of a sector of the unit circle. Let us confine our attention to the first quadrant, in which a quarter of the unit circle is represented by $y = \sqrt{1-x^2}$, $0 \leq x \leq 1$. For reasons to be explained, we want to avoid the neighborhood of $x = 1$ and limit ourselves, therefore, to $0 \leq x \leq \frac{1}{2}$. The area under the graph of $y = \sqrt{1-x^2}$ over this interval is denoted by A. Figure 5.3 shows that $A = S + T$, where S is the area of a sector with a vertex angle of 30°, and hence $S = \pi/12$. Also, T is the area of a 30°–60°–90° triangle, and hence $T = \sqrt{3}/8$. We therefore have $A = \sqrt{3}/8 + \pi/12$, or

$$\pi = 12A - \frac{3}{2}\sqrt{3} \tag{5.15}$$

so the computation of π will have been accomplished once we have computed the area A.

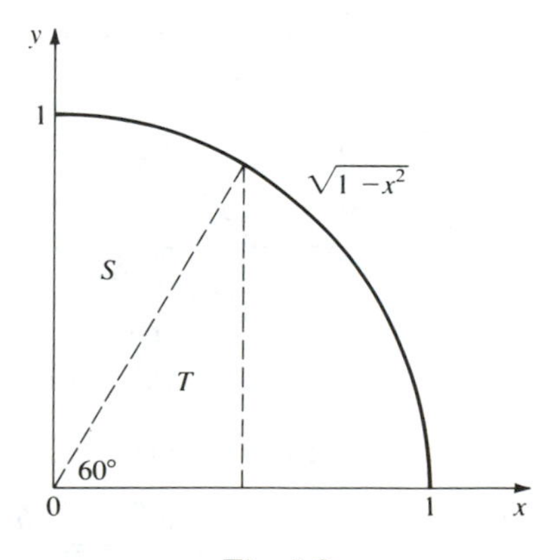

Fig. 5.3

To compute A to a high accuracy, let us partition the interval $0 \le x \le \frac{1}{2}$ into n equal subintervals of width $h = 1/(2n)$ and carry out the area computations detailed in Chapter 3. We also quote from Chapter 3 the following two results for error estimates associated with the computation of the area under the graph of $f(x)$ for $a \le x \le b$.

Rectangular approximation. If the area under a monotone function is approximated by the rectangular method, then the error incurred does not exceed C/n, where $C = (b-a)|f(b) - f(a)|$.

Trapezoidal approximation. If the area under a convex (concave) function is approximated by the trapezoidal method, then the error incurred does not exceed D/n^2, where D is independent of n (and h) and satisfies

$$D \ge K = \frac{(b-a)^2}{2}\left|\frac{f(b+h)-f(b)}{h} - \frac{f(a+h)-f(a)}{h}\right| \tag{5.16}$$

In our case $a = 0$, $b = \frac{1}{2}$, $f(a) = 1$, $f(b) = \sqrt{3}/2$, so

$$C = \frac{1}{2}\left|\frac{\sqrt{3}}{2} - 1\right| = \frac{2-\sqrt{3}}{2} < \frac{1}{12} \tag{5.17}$$

Accordingly, the error in the rectangular approximation of A in (5.15) is less than $1/(12n)$, and because A is multiplied by 12, so is the error. Thus, the error incurred in approximating π through (5.15), using the rectangular method, is less than $1/n$.

Next we consider the trapezoidal approximation and examine (5.16) for $f(x) = \sqrt{1-x^2}$. After rationalizing, we have

$$\frac{f(a+h)-f(a)}{h} = \frac{\sqrt{1-h^2}-1}{h} = \frac{-h}{\sqrt{1+h^2}+1} \tag{5.18}$$

and

$$\begin{aligned}\frac{f(b+h)-f(b)}{h} &= \frac{\sqrt{1-\left(\frac{1}{2}+h\right)^2} - \sqrt{1-\left(\frac{1}{2}\right)^2}}{h} \\ &= \frac{-2(1+h)}{\sqrt{3}+\sqrt{3-4h-4h^2}}\end{aligned} \tag{5.19}$$

Accordingly, the error bound in (5.16) appears as

$$K = \frac{1}{8}\left|\frac{2(1+h)}{\sqrt{3}+\sqrt{3-4h-4h^2}} - \frac{h}{\sqrt{1+h^2}+1}\right| \tag{5.20}$$

The negative term in (5.20), for the relevant values of $h \leq \frac{1}{2}$, can be omitted (see Exercise 5.8) to give

$$K \leq \frac{\frac{1+h}{4}}{\sqrt{3}+\sqrt{3-4h-4h^2}} < \frac{\frac{1+h}{4}}{\sqrt{3}} < \frac{1+h}{6} \leq \frac{1}{4} \tag{5.21}$$

Thus we can choose $D = \frac{1}{4}$, and, accordingly, the error incurred in approximating π through (5.15) with the trapezoidal method is less than $12/(4n^2) = 3/n^2$. Because the essential feature of this error bound is its denominator n^2, we can afford to be generous in obtaining an upper bound for K in (5.20) and use $D = \frac{1}{4}$.

Suppose we want to compute π with an error not exceeding $\frac{1}{2} \cdot 10^{-q}$; that is, we want to obtain q correct decimal figures. The rectangular method requires $1/n \leq (1/2) \cdot 10^{-q}$, and thus $n = 2 \cdot 10^q$ suffices. On the other hand, the trapezoidal method requires only $3/n^2 \leq (1/2) \cdot 10^{-q}$, so $n = (2.5) \cdot 10^{q/2}$ will do. For $q = 5$, say, the rectanglar method necessitates $n = 200,000$ whereas the more efficient trapezoidal method delivers the same accuracy using $n = 800$. These computations have been carried out, yielding the approximation 3.14159 in both cases.

As explained in Chapter 3, the rectangular method is inefficient. It is therefore suggested that the laboratory participants experiment with the trapezoidal method as we have outlined for the computation of π to various degrees of accuracy. In this connection see Exercise 5.9.

5.5 Buffon's needle

In this section we offer laboratory participants a chance to tackle the so-called Buffon's needle method (named after Count Buffon who discovered it in 1777). Unlike the previous two methods, this method calls for some rudimentary knowledge of integrals.

Suppose we toss a large number of identical needles (matches, toothpicks) into the air. The needles fall onto a large striped sheet, whose uniform stripe width W equals (for simplicity) the length of the needles. The needles could also be tossed up one at a time provided the tosses are random, without any preferred direction or initial position (one single needle may then suffice). A typical pattern after the needles land is shown in Fig. 5.4.

We want to investigate the probability that a landing needle will cross the common boundary of two stripes. An entertaining description of Buffon's experiment is given by George Gamow (in *One Two Three*

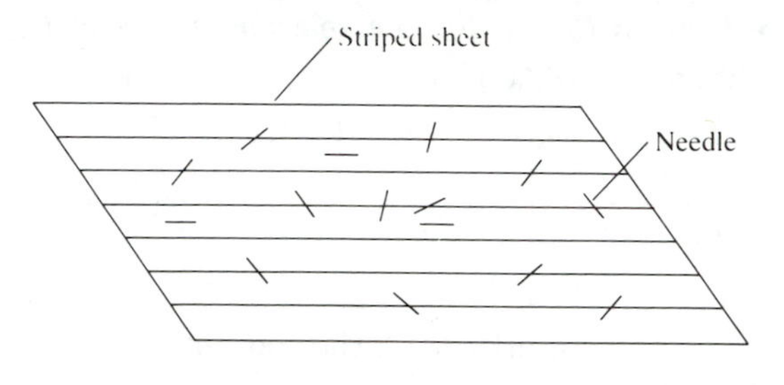

Fig. 5.4

Infinity, pp. 210–213), who termed it "Finding π by throwing matches on the American flag." Clearly, some of the needles will be completely inside a stripe whereas others will lie across a boundary. Before we investigate the probability that a randomly tossed needle will land one way or the other, we suggest that the laboratory participants be familiarized with the notion of probability in an intuitive way, using a host of elementary examples (dice, coins, cards). In particular, they should realize that the probability P of an event occurring (such as a needle lying across a boundary) satisfies $0 \leq P \leq 1$. In this connection see Exercises 5.10 and 5.11.

Returning to our problem of the landing position of the tossed needles, let us characterize the position of a needle with respect to the stripe containing the needle's midpoint by assigning to the needle a pair of numbers D and α. The number D is the distance of the needle's midpoint from the nearest boundary, and α is the acute angle formed by the needle and the direction of the stripes. Clearly, $0 \leq D \leq W/2$ and $0 \leq \alpha \leq \pi/2$ (radians). The numbers D and α are independent random numbers because the tosses are assumed to have been random. The situation is shown in Fig. 5.5, where the projection of the needle's semilength on the direction perpendicular to the stripes is given by $(W/2)\sin\alpha$. If $D < (W/2)\sin\alpha$, crossing takes place, as seen in Fig. 5.5, whereas no crossing occurs if the opposite is true (we recall that the length of the needles equals the stripe width W). The laboratory instructor should explain that the case $D = (W/2)\sin\alpha$ can be considered to belong to either crossing or no crossing without affecting the results.

The totality of positions that a needle can assume can be represented by the totality of points of the rectangle shown in Fig. 5.6. The abscissa represents the angle α, which ranges from 0 to $\pi/2$, whereas the ordinate D ranges from 0 to $W/2$. The totality of points in the rectangle representing positions of the crossing needles is therefore given by the

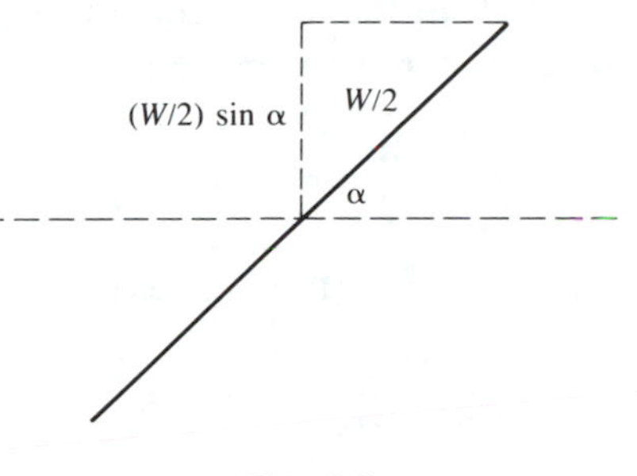

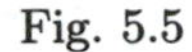
Fig. 5.5

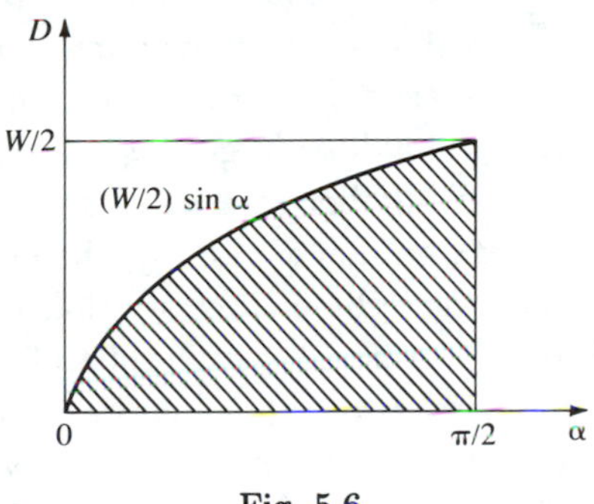

Fig. 5.6

totality of points satisfying $D < (W/2)\sin\alpha$; these points comprise the shaded area of the rectangle in Fig. 5.6. The blank area of the rectangle represents all positions of noncrossing needles.

Now that we know which fraction of the rectangle represents all possibilities of crossings, we can express the probability that a randomly tossed needle will lie across a boundary as the fraction whose numerator is the shaded area in Fig. 5.6 and whose denominator is the area of the entire rectangle. This ratio is obviously a number between zero and one (as are all probabilities) and will be calculated next. That this ratio indeed equals the probability we are after can be seen by analogy to the die problem. There is exactly one way to obtain the number 3, say, when rolling a die, and there are six possible outcomes of a roll. Hence, the probability of obtaining the number 3 is one out of six, or $\frac{1}{6}$. Similarly, there are three ways of obtaining an even number when rolling a die (2, 4, or 6), so the probability of getting an even number is three out of six, or the ratio $\frac{3}{6} = \frac{1}{2}$.

The area of the rectangle in Fig. 5.6 is equal to $(\pi/2) \cdot (W/2)$. On the other hand, the area of the shaded part equals

$$\int_0^{\pi/2} \frac{W}{2} \sin\alpha \, d\alpha = \frac{W}{2} \tag{5.22}$$

Consequently, the required probability equals

$$\frac{\text{shaded area}}{\text{entire area}} = \frac{1}{\frac{\pi}{2}} = \frac{2}{\pi} \tag{5.23}$$

Now suppose we toss a very large number of needles and count how many of them cross one of the boundaries. If we divide that number by the number of all the needles tossed, this ratio will be close to $2/\pi$. The reason we insist on tossing a very large number of needles is that a small number of tosses give less accurate results. Imagine rolling a die six times. Although the probability of getting the number 3, say, is $\frac{1}{6}$, we usually do not get that number in exactly one of these six throws. However, if we roll a die six million times, say, the number 3 will indeed appear in approximately one million throws – that is, in approximately one sixth of the throws, consistent with probability. In other words, probability predicts the number of times a certain event will happen for a very large number of trials. In like manner, when we toss a very large number T of needles, for which C crossings occur, then

$$P = \lim_{T\to\infty} \frac{\text{number of crossings } C}{\text{number of tossed needles } T} = \frac{\text{shaded area}}{\text{entire area}} = \frac{2}{\pi} \tag{5.24}$$

The intuitive notion of probability that led us to (5.24) implies that for large T the quantity $2T/C$ should be a fair approximation to π. We also expect the quality of this approximation to improve with increasing T (though not monotonically).

At this point a remark is in order. Had we defined α to be the angle formed by the needle and the positive direction of the stripes, we would have had $0 \leq \alpha < \pi$ (the two ends of the needle are indistinguishable). Following the arguments leading to Fig. 5.6, we would obtain the same figure but augmented by its reflection across the line $\alpha = \pi/2$. Thus, the shaded area and the area of the rectangle in Fig. 5.6 would both double, and (5.23) and (5.24) remain valid.

The experiment just described should be carried out in the mathematical laboratory, but it turns out that an extremely large number of tosses is required just to obtain the modest approximation 3.14. Fortunately, we can harness the computer to *simulate* the tossings for us. The laboratory instructor should dwell on the concept of *simulation*, its implementation, and its far-reaching benefits. For our particular purpose (see Chapter 8), suppose our computer has a built-in function $SIND(A)$, which gives a highly accurate value of the sine of any angle A, expressed in degrees. We also assume that we have a built-in random-number generator $RAND(x)$, which produces a random number r such that $0 \leq r < x$, all numbers in this range having an equal chance of being picked. (Various subtleties associated with computer-generated random numbers are glossed over; also see Exercises 5.12 and 5.13.) We have chosen to express A in degrees, although the preceding analysis was in terms of radians, because of the integration in (5.22). The reason is that we should not instruct the computer to generate a random number between zero and $\pi/2$ in the very process of trying to approximate the value of π.

We next proceed to simulate the Buffon experiment by means of an algorithm. Without loss of generality we can set $W = 2$ because the ratio of the areas in (5.23) is independent of W. Consequently, the semilength $W/2$ of the needles equals unity. It follows that the distance D is a random number in $[0, 1)$, and the angle A is a random number in $[0, 90)$. Now we can write an algorithm that, instead of each toss of a needle, generates the pair (D, A) of random numbers. The cases for which $D < \sin A$ correspond to the cases of needles lying across a boundary and are therefore counted. Forming the ratio of their number to the total number of cases, we obtain an approximation for $2/\pi$ as before. Moreover, the enormous speed of the computer enables us to perform these simulated tosses a vast number of times rather quickly.

The following algorithm can be used to obtain an approximation for π by the simulation of Buffon's experiment:

1. Input the maximum number M of tosses (tens of thousands).
2. Set $T = 0$ and $C = 0$.
3. Increase T by 1.
4. Compute $D = RAND(1)$.
5. Compute $A = RAND(90)$ and $S = SIND(A)$.
6. If $D < S$, then increase C by 1.
7. It T is divisible by 10000, then print T, C, and $(2T/C)$.
8. If $T < M$, then return to Step 3.
9. End.

We actually ran this algorithm with $M = 10^7$. A few representative values obtained are reproduced in Table 5.4.

Table 5.4. *Results of Buffon simulation*

T	C	$2T/C$
10000	6345	3.1521
100000	63731	3.1382
500000	318470	3.1400
1000000	636759	3.1409
10000000	6366183	3.1416

As mentioned, the subtleties inherent in the use of random-number generators (especially for a vast number of times) are left in the background. A close examination shows that if our only interest were to calculate π with ever-increasing accuracy, we would end up testing the quality of the generator $RAND(x)$ rather than approximating π. Moreover, Table 5.4 shows the convergence to be deplorably slow. Nevertheless, we recommend including the Buffon method among the assignments of the mathematical laboratory. It has a considerable mathematical and educational value in that it introduces *probabilistic methods* into the realm of computational mathematics, and those methods lead to the Monte-Carlo methods (see Exercises 5.14 and 5.15). In addition, the laboratory participants are given the opportunity to become acquainted with the concept of *simulation*.

In his book, G. Gamow tells about the Italian mathematician Lazzerini who carried out the Buffon experiment by tossing 3407 matches, of which 2169 crossed boundaries. Thus, Lazzerini obtained the fantastic approximation 3.14154, which is correct to four decimal figures. Such accuracy is *not* characteristic of the Buffon method, whose expected error can be shown (by advanced methods) to behave like $1/\sqrt{T}$. Lazzerini's result is therefore but a lucky strike, and laboratory participants must not expect such accuracy even for millions of tosses. Table 5.4, on the other hand, exhibits a typical outcome of a Buffon experiment (see Exercises 5.16 and 5.17).

5.6 The arctangent method

The method to be described here is based on the use of a geometric series for the expansion of $\arctan x$ in a power series and is intended therefore for cocalculus-level laboratory participants. Those who have had no calculus at all may take the expansion (5.27) on faith or else skip this section without loss of continuity. Let us start from the elementary geometric series

$$\frac{1}{1+t^2} = \frac{1}{1-(-t^2)} = 1 - t^2 + t^4 - + \cdots + (-1)^{k-1} t^{2k-2} + R \quad (5.25)$$

where

$$R = (-1)^k [t^{2k} - t^{2k+2} + \cdots] = \frac{(-1)^k t^{2k}}{1+t_2}, \qquad |t| < 1 \quad (5.26)$$

Next, we integrate both sides of (5.25) from $t = 0$ to $t = x$, to obtain

$$\arctan x = x - \frac{x^3}{3} + \frac{x^5}{5} - + \cdots + (-1)^{k-1} \frac{x^{2k-1}}{2k-1} + E \quad (5.27)$$

where

$$E = (-1)^k \int_0^x \frac{t^{2k}}{1+t^2}\, dt \quad (5.28)$$

The expression E is an error term that can be estimated by

$$|E| = \int_0^x \frac{t^{2k}}{1+t^2}\, dt \leq \int_0^x t^{2k}\, dt = \frac{x^{2k+1}}{2k+1} \quad (5.29)$$

and we note that for $|x| \leq 1$, $|E|$ approaches zero as $k \to \infty$. Recalling that $\arctan(1) = \pi/4$, we are tempted to compute π by setting $x = 1$ and obtain the Gregory–Leibniz series (discovered in 1671)

$$\frac{\pi}{4} = 1 - \frac{1}{3} + \frac{1}{5} - + \cdots + \frac{(-1)^{k-1}}{2k-1} + E \quad (5.30)$$

with $|E| \leq 1/(2k+1)$. However, if we require an error not exceeding $\frac{1}{2} \cdot 10^{-q}$, we find that $k = 10^q$, that is, a million terms are needed for, say, six decimal figures. This is clearly unacceptable from a practical point of view.

It is still possible, however, to use (5.27) efficiently by employing an idea attributed to the English mathematician John Machin (who suggested its use in 1706). If we set $u = \tan\alpha$ and $v = \tan\beta$ in the addition formula for the tangent

$$\tan(\alpha + \beta) = \frac{\tan\alpha + \tan\beta}{1 - \tan\alpha \tan\beta} \quad (5.31)$$

and take the arctangent of both sides, we obtain

$$\arctan u + \arctan v = \arctan \frac{u+v}{1 - u \cdot v} \quad (5.32)$$

Choosing, for example, $u = \frac{1}{2}$, $v = \frac{1}{3}$, we find $(u+v)/(1-uv) = 1$, and thus

$$\frac{\pi}{4} = \arctan\frac{1}{2} + \arctan\frac{1}{3} \tag{5.33}$$

The right-hand side of (5.33) can be obtained from (5.27) for $x = \frac{1}{2}$ and $x = \frac{1}{3}$, respectively, and for simplicity we take an equal number of terms in both series. We thus obtain

$$\begin{aligned}\frac{5.45\pi}{4} = &\left[\frac{1}{2} - \frac{\left(\frac{1}{2}\right)^3}{3} + \frac{\left(\frac{1}{2}\right)^5}{5} - + \cdots + (-1)^{k-1}\frac{\left(\frac{1}{2}\right)^{2k-1}}{2k-1}\right] \\ &+ \left[\frac{1}{3} - \frac{\left(\frac{1}{3}\right)^3}{3} + \frac{\left(\frac{1}{3}\right)^5}{5} - + \cdots + (-1)^{k-1}\frac{\left(\frac{1}{3}\right)^{2k-1}}{2k-1}\right] \\ &+ E_{1/2} + E_{1/3}\end{aligned} \tag{5.34}$$

where

$$|E_{1/2} + E_{1/3}| \le \frac{\left(\frac{1}{2}\right)^{2k+1} + \left(\frac{1}{3}\right)^{2k+1}}{2k+1} < \frac{2}{2k+1}\left(\frac{1}{2}\right)^{2k+1} \tag{5.35}$$

As before, we require the total error not to exceed $\frac{1}{2} \cdot 10^{-q}$, but now we find that we must have $10^q \le (k+0.5)2^{2k}$; so k satisfying $10^q \le 2^{2k}$ will certainly do. Now, $2^{10} > 10^3$ so an easy calculation shows that $k \ge \frac{5}{3}q$ is more than enough to meet our requirements. Choosing $q = 6$, we find that ten terms supplies us with the six decimal figures for which we needed a million terms using (5.30). The instructor should point out to the laboratory participants that the efficiency of the method hinges on the fact that (5.27) is used with $|x| < 1$, so its terms decrease rapidly with increasing k. For the very same reason, even greater efficiency is attained by repeated use of (5.32) for various appropriate values of u and v. Thus, we find $\arctan\frac{1}{3} = \arctan\frac{1}{5} + \arctan\frac{1}{8}$ and $\arctan\frac{1}{2} = \arctan\frac{1}{3} + \arctan\frac{1}{7}$, which leads (see Exercise 5.18) to

$$\frac{\pi}{4} = 2\arctan\frac{1}{5} + \arctan\frac{1}{7} + 2\arctan\frac{1}{8} \tag{5.36}$$

An analysis similar to the preceding one shows (see Exercise 5.19) that $k \geq \frac{5}{7}q$ guarantees q decimal figures, so six-digit accuracy can be attained with five terms.

The process can be continued along similar lines, and we can obtain formulas for $\pi/4$ such as

$$\frac{\pi}{4} = 6 \arctan \frac{1}{8} + 2 \arctan \frac{1}{57} + \arctan \frac{1}{239} \tag{5.37}$$

and

$$\frac{\pi}{4} = 12 \arctan \frac{1}{18} + 8 \arctan \frac{1}{57} - 5 \arctan \frac{1}{239} \tag{5.38}$$

and so on. These formulas have actually been used for computer calculations of π to exceedingly high accuracy. Thus, in 1961 Shanks and Wrench employed them to compute π to 100,000 decimal digits, using about eight hours on an IBM-704 computer in New York. Six years later, Gilloud and Filliatre used a CDC-6600 computer in Paris, which ran twenty-eight hours to obtain 500,000 decimal digits. Had they used the IBM-704 computer, it would have taken them hundreds of hours. In 1972, one of the authors' students, A. Wandel, used the very same formulas on Tel Aviv University's CDC-6600 and obtained 3000 digits in five minutes. This result was obtained in the framework of the authors' first mathematical laboratory for students. In 1973, Gilloud and Bouyer passed the million-digit mark, a feat that took just under a day of computation on a CDC-7600.

It is clear that such computations are intended mainly to demonstrate the ever-increasing power of computer technology and advanced programming techniques rather than to obtain hundreds of thousands of decimal digits for π. They also demonstrate how computational efficiency can be brought about by using convergence-accelerating techniques. Thus, the laboratory participants can be impressed with the importance of the speed of convergence in a practical computation, beyond the theoretical proof that the process converges (see Chapter 6). Let them use (5.27) and (5.32), choose judicious values of u and v, obtain their own homemade formulas, and run them on their microcomputers (see Exercises 5.20 and 5.21). The following typical algorithm for (5.34) is suggested.

1. Input $q =$ desired number of correct figures.
2. Set $k =$ the first integer not less than $\frac{5}{3}q$.
3. Set $u = \frac{1}{2}$, $v = \frac{1}{3}$, $s = 0$, $n = 1$, and $m = 2k - 1$.
4. Replace s by $s + (u + v)/n$.
5. Print the value of $4s$.

Table 5.5. *Arctangent approximations of* π

Number of Terms	Approximate Value of π
1	3.33333333333333
2	3.11728395061728
3	3.14557613168724
4	3.14085056176106
5	3.14174119743369
6	3.14156158787759
7	3.14159934096620
8	3.14159118436091
9	3.14159298133457
10	3.14159257960635
11	3.14159267045069
12	3.14159264971679
13	3.14159265448535
14	3.14159265338154
15	3.14159265363846
16	3.14159265357837
17	3.14159265359248
18	3.14159265358916
19	3.14159265358994
20	3.14159265358976
21	3.14159265358980
22	3.14159265358979
23	3.14159265358979
24	3.14159265358979

6. Replace u by $(-u/4)$ and v by $(-v/9)$.
7. If $n < m$, then replace n by $(n + 2)$ and return to Step 4.
8. End.

The algorithm avoids the use of powers of u, v, and (-1) but uses Step 6 to generate the next required numerators. Moreover, Step 5 prints the current approximation for π so that the laboratory participants can see the ever-improving results in the making. We actually ran a program based on this algorithm and obtained the results shown in Table 5.5 in the next section. The computations were carried out with double precision on a PC and terminated in accordance with $q = 14$ in Step 1 of the algorithm. It is seen that 14-digit accuracy is already obtained after 22 terms. Nevertheless, the program used 24 terms, owing to the generous estimates in (5.35) and the discussion thereafter, which dictated a slightly exaggerated number of terms.

5.7 A brief historical sketch

The quest for ever-increasing accuracy in the computation of π is at least as ancient as the Bible. So far we have discussed four methods for the computation of π that are suitable for the mathematical laboratory. Two of those presupposed no knowledge of calculus, one required elementary integration, and the fourth touched on power-series expansions. Many other methods are available, of course, some of which could also be implemented in the mathematical laboratory. We chose these four methods because of their educational value, in that they present a variety of ideas, both old and new, and broaden the mathematical background of the laboratory participants. If we were to add a fifth method, we would choose Wallis' product or Euler's continued-fraction representation. Each of these methods sheds additional light on the computation of π, from different points of view.

Here is a brief historical sketch of the computation of π (for more details, see, for example, P. Beckmann, *A History of* π):

King Solomon (*c.* 975 B.C.)	3.0
Archimedes (*c.* 240 B.C.)	3.14
Aryabhata (*c.* 450 A.D.)	3.1416
Vieta (1593)	3.141592653
Newton (1666)	3.1415926535897932
Machin (1706)	3.(100 digits)
Vega (1794)	3.(140 digits)
Dase & Strassnitzky (1844)	3.(200 digits)
Ferguson (1947)	3.(800 digits)
Reitwiesner (1949)	3.(2000 digits)
Shanks & Wrench (1962)	3.(100000 digits)
Gilloud & Filliatre (1968)	3.(500000 digits)
Gilloud & Bouyer (1973)	3.(one million digits)
Kanada (1987)	3.(134 million digits)

Dase and Strassnitzky obtained their results by long and tedious hand computations whereas Ferguson used a desk calculator. Reitwiesner followed John von Neumann's suggestion and programmed ENIAC, one of the first computers, to arrive at his value for π. Indeed, his calculations (in 1949) were completed over the Labor Day weekend, when he and three of his colleagues at Aberdeen Research Laboratories took 8-hour shifts to keep ENIAC operating continuously. Shanks and Wrench in New York, as well as Gilloud, Filliatre, and Bouyer in Paris, still employed arctangent methods to arrive at their results. It soon became clear, however, that these methods were subject to inescapable limits, even if computing power and speed increased a hundredfold. For example, Gilloud and Bouyer's program would have required at least a quarter of a century to produce a billion-digit value for π. Thus, all subsequent digit hunters used different methods, which are iterative and at least quadratically convergent.

The results of Kanada (Tokyo, 1987) were obtained on an NEC-SX2 supercomputer, using the Gauss–Brent–Salamin iterative algorithm. This algorithm is based on the work of Srinivasa Ramanujan (1914) and related efforts of J. Borwein and P. Borwein (1986), who state that if one could somehow monopolize a supercomputer for a few weeks, more than two billion digits of π could be produced!

5.8 Compound interest and the number e

Another number that plays a central role in many branches of mathematics and is suitable for a numerical laboratory assignment is the number e, which surfaces naturally in modern life. We do not have to look far to find illustrative examples: interest earned continuously on a bank deposit, depreciation of equipment, and growth and decay processes in biology and medicine.

A simple introduction to the number e is the analysis of compound interest. Suppose we deposit an amount of A_0 dollars in a bank at a rate of p percent per annum (e.g., 6%), compounded annually. At the end of one year the amount A_0 will have grown to

$$A_1 = A_0 + A_0 \frac{p}{100} = A_0 \left(1 + \frac{p}{100}\right) \tag{5.39}$$

Similarly, after two years the initial amount A_0 will have grown to

$$A_2 = A_1 + A_1 \frac{p}{100} = A_1 \left(1 + \frac{p}{100}\right) = A_0 \left(1 + \frac{p}{100}\right)^2 \tag{5.40}$$

After n years the money on deposit is

$$A_n = A_0 \left(1 + \frac{p}{100}\right)^n \tag{5.41}$$

For example, if $A_0 = 1000$ and $p = 6$, then after 10 years the initial deposit A_0 will have grown to $A_{10} = 1000(1.06)^{10}$, or $A_{10} = 1790.85$, correct to two decimal figures.

Let us now consider the following question: If, instead of compounding interest annually at 6 percent per annum, the bank offers to compound the same interest semiannually – 3 percent every 6 months – would the depositor gain or lose?

At first sight, 6 percent per year appears to be equivalent to 3 percent every half-year. However, a little reflection shows that the bank's offer was in the depositor's favor. Clearly, 3 percent of interest has accumulated by the end of 6 months and forms, together with the initial deposit, a larger sum for the next 6 months. In other words, at the end of one year the total interest accumulated will be more than 6 percent of the initial deposit. Specifically, after one year the deposit A_0 will have grown to $A_0\left(1+\frac{3}{100}\right)^2 = A_0(1.0609)$; that is, 6 percent interest compounded semiannually is effectively equivalent to 6.09 percent compounded annually.

Let $A_{n,k}$ denote the amount accumulated after n years when A_0 dollars are deposited and interest is compounded k times annually. Accordingly, as we saw for $A_0 = 1000$ and $p = 6$,

$$A_{10,1} = 1000\left(1+\frac{6}{100}\right)^{10} = 1790.85 \tag{5.42}$$

correct to two decimal figures, whereas

$$A_{10,2} = 1000\left(1+\frac{3}{100}\right)^{20} = 1806.11 \tag{5.43}$$

because 10 years contain 20 half-year periods.

Suppose next that interest is compounded quarterly (i.e., 1.5% every 3 months). Then

$$A_{10,4} = 1000\left(1+\frac{1.5}{100}\right)^{40} = 1814.02 \tag{5.44}$$

We see that the depositor gains again from the more frequent compounding of interest, but the gains decrease in magnitude: $A_{10,2}-A_{10,1} = 15.26$ whereas $A_{10,4} - A_{10,2} = 7.91$.

More generally, if interest of p percent per annum is compounded k times annually (p/k percent every kth of a year), then after n years we have

$$A_{n,k} = A_0\left(1+\frac{\frac{p}{k}}{100}\right)^{kn} \tag{5.45}$$

The question now arises as to whether we can turn an initial deposit into an arbitrarily large sum by compounding interest sufficiently frequently, or whether there is a limit to the amount we can generate from a given initial deposit by frequent compounding of interest.

Before we turn to the answer of this question, it is worthwhile pointing out that the question is by no means merely academic. Indeed, there are quite a few "natural banks" that compound interest every instant, that is, continuously. That is to say, many natural phenomena (some will be discussed later) behave according to the compound interest law. Obviously, nature does not behave in annual jumps, say once a year on New Year's Day, nor does it carry a wristwatch telling it to act every hour on the hour. Rather, nature follows its course smoothly and continuously in the sense of compounding interest k times a year with k becoming infinitely large.

With a view toward answering the question we have raised, let us simplify the notation by defining $p/100 = \alpha$. Then (5.45) can be rewritten in the form

$$A_{n,k} = A_0 \left(1 + \frac{\alpha}{k}\right)^{kn} \tag{5.46}$$

We divide both sides by A_0 and take the nth root to obtain

$$\left(\frac{A_{n,k}}{A_0}\right)^{1/n} = \left(1 + \frac{\alpha}{k}\right)^{k} \tag{5.47}$$

where, of course, $x^{1/n} = \sqrt[n]{x}$. We recall that A_0, being the initial deposit, is a fixed quantity, as is the number n of years. Therefore, the behavior of $A_{n,k}$ as k becomes infinitely large depends entirely on the behavior, as k grows, of the right side of (5.47).

To study the behavior of $(1 + \alpha/k)^k$ as k increases, let us first take the simple case $\alpha = 1$ (i.e., $p = 100$). The general case will be discussed later. To facilitate the ensuing analysis, we also set

$$a_k = \left(1 + \frac{1}{k}\right)^{k} \tag{5.48}$$

Let us take a close look at the first 10 values $a_1, a_2, \ldots, a_{10}$, as well as the values of a_k for several higher values of k. We tabulate them in Table 5.6, which displays 4-decimal figures.

A look at the table reveals that a_k increases as k increases, but the increments keep getting smaller. Moreover, it seems very unlikely that a_k will ever exceed the value 3, however large k might be. It is these remarks that we are about to prove rigorously.

Table 5.6. *Values of a_k*

k	$a_k = \left(1+\frac{1}{k}\right)^k$
1	2.0000
2	2.2500
3	2.3703
4	2.4414
5	2.4883
6	2.5216
7	2.5464
8	2.5658
9	2.5812
10	2.5937
⋮	⋮
32	2.6770
⋮	⋮
64	2.6973

Recall the binomial expansion

$$(x+y)^m = x^m + \binom{m}{1}x^{m-1}y + \binom{m}{2}x^{m-2}y^2 + \cdots + \binom{m}{m-1}xy^{m-1} + y^m \tag{5.49}$$

in which we have $(m+1)$ terms, each of which consists of a binomial coefficient multiplying a product of a power of x by a power of y. The exponents of x and y in each term add up to m. The binomial coefficient is defined by

$$\binom{m}{r} = \frac{m!}{r!(m-r)!} \tag{5.50}$$

where $r! = 1 \cdot 2 \cdot 3 \cdot \cdots \cdot r$ and $0! = 1$. A little reflection (see Exercise 5.22) shows that

$$\begin{aligned} \binom{m}{r} &= \binom{m}{m-r} = \frac{m(m-1)(m-2)\cdots(m-r+1)}{r!} \\ \binom{m}{0} &= \binom{m}{m} = 1 \\ \binom{m}{1} &= \binom{m}{m-1} = m \end{aligned} \tag{5.51}$$

Now let us set $x = 1$, $y = 1/k$, and $m = k$ in (5.49). Then the left side becomes a_k, defined in (5.48), and we have

$$a_k = 1 + \binom{k}{1}\frac{1}{k} + \binom{k}{2}\left(\frac{1}{k}\right)^2 + \binom{k}{3}\left(\frac{1}{k}\right)^3 + \cdots + \binom{k}{k}\left(\frac{1}{k}\right)^k \tag{5.52}$$

Using (5.50), we can cast a_k into the form

$$\begin{aligned} a_k &= 1 + k\frac{1}{k} + \frac{k(k-1)}{2!}\frac{1}{k^2} + \cdots + \frac{k(k-1)\cdots 1}{k!}\frac{1}{k^k} \\ &= 1 + 1 + \frac{1}{2!}\left(1 - \frac{1}{k}\right) \\ &\quad + \cdots + \frac{1}{k!}\left(1 - \frac{1}{k}\right)\left(1 - \frac{2}{k}\right)\cdots\left(1 - \frac{k-1}{k}\right) \end{aligned} \tag{5.53}$$

From (5.53) we infer (see Exercise 5.23) that a_k is an increasing sequence, as expected; that is, $a_{k+1} > a_k$. Moreover, because $a_1 = 2$, we have $a_k > 2$ for all $k > 1$.

As mentioned, it appears from Table 5.6 that a_k is not likely to exceed the value 3 however large k might be. We now prove that such is indeed the case. If we take another look at a_k in (5.53), we find that the typical term, except the first two, is given by

$$\frac{1}{r!}\left(1 - \frac{1}{k}\right)\left(1 - \frac{2}{k}\right)\cdots\left(1 - \frac{r-1}{k}\right) \le \frac{1}{r!} \tag{5.54}$$

because none of the factors in parenthesis exceeds the value 1. Consequently,

$$a_k \ge 1 + 1 + \frac{1}{2!} + \frac{1}{3!} + \cdots + \frac{1}{k!} \tag{5.55}$$

On the other hand, it is easy to see that

$$\begin{aligned} \frac{1}{3!} &= \frac{1}{1\cdot 2\cdot 3} < \frac{1}{2\cdot 2} = \left(\frac{1}{2}\right)^2 \\ \frac{1}{4!} &= \frac{1}{1\cdot 2\cdot 3\cdot 4} < \frac{1}{2\cdot 2\cdot 2} = \left(\frac{1}{2}\right)^3 \end{aligned} \tag{5.56}$$

and, in general,

$$\frac{1}{r!} = \frac{1}{1\cdot 2\cdot 3\cdot 4\cdot\;\cdots\;\cdot r} < \left(\frac{1}{2}\right)^{r-1} \tag{5.57}$$

for all integral values of $r > 1$. Using (5.55)–(5.57), we obtain

$$a_k < 1 + \left[1 + \frac{1}{2} + \left(\frac{1}{2}\right)^2 + \cdots + \left(\frac{1}{2}\right)^{k-1}\right] \tag{5.58}$$

Note that the expression inside the square brackets is a geometric progression. For a geometric progression, we have

$$c + cq + cq^2 + \cdots + cq^{k-1} = c\frac{1-q^k}{1-q} \tag{5.59}$$

Here $c = 1$ and $q = \frac{1}{2}$. Therefore,

$$a_k < 1 + \frac{1 - \left(\frac{1}{2}\right)^k}{1 - \frac{1}{2}} = 1 + 2\left[1 - \left(\frac{1}{2}\right)^k\right] < 3 \tag{5.60}$$

Collecting our results, we find that $a_k = (1 + 1/k)^k$ is an increasing sequence, each of whose members is not less than two but less than three, that is,

$$2 \leq a_k = \left(1 + \frac{1}{k}\right)^k < 3 \tag{5.61}$$

A sequence such as a_k, whose values are constrained to lie between two fixed numbers, is said to be a *bounded sequence*. Thus, a_k is an increasing bounded sequence. The laboratory participants are called on to use an alternative, elegant method to arrive at the same result with the aid of the arithmetic–geometric inequality. In this connection see Exercises 5.24 and 5.25.

It is shown in calculus that an increasing (or, for that matter, decreasing) bounded sequence tends to a limiting value. This limiting value, also called the *limit*, is the number that the terms a_k of the sequence approach as they increase (decrease) with increasing k. Moreover, this number is approached arbitrarily closely as k becomes larger. A formal, rigorous definition of the limit concept is given in the framework of calculus, but the laboratory participants should make do, for the time being, with the preceding intuitive description.

The limit of our sequence turns out to play a central role in many areas of mathematics and the sciences, such as engineering, biology, physics, and economics. Because of its importance, this limit is designated by the special symbol e. (This is akin to the symbol π.) We shall see later how to compute e to any desired accuracy. The symbols e and π, incidentally, were introduced by Leonhard Euler (1707–1783).

We express the fact that the sequence a_k approaches e as k increases indefinitely, by writing

$$\left(1+\frac{1}{k}\right)^k \longrightarrow e \tag{5.62}$$

In calculus it is shown, analogously, that

$$\left(1+\frac{\alpha}{k}\right)^k \longrightarrow e^\alpha \tag{5.63}$$

for any number α. Clearly, when $\alpha = 1$, (5.63) reduces to (5.62).

Returning to our point of departure, we recall that the amount to which an initial deposit A_0 grows after n years, at p percent per annum compounded k times annually, is given from (5.45) by

$$A_{n,k} = A_0\left[\left(1+\frac{\alpha}{k}\right)^k\right]^n, \qquad \alpha = \frac{p}{100} \tag{5.64}$$

As k grows indefinitely (continuous compounding of interest), (5.64) in conjunction with (5.63) show that

$$A_{n,k} \longrightarrow A_0 e^{\alpha n} \tag{5.65}$$

Let us now denote by $\widetilde{A}_n$ the amount to which an initial deposit of A_0 grows, at p percent per annum, after n years of continuous compounding of the interest. Then,

$$\widetilde{A}_n = A_0 e^{\alpha n} \tag{5.66}$$

We show the import of (5.66) by a few examples, which we touched on earlier. Suppose we deposit 1000 dollars for 10 years at 6 percent per annum compounded continuously. The 1000 dollars will then grow according to (5.66) to

$$\widetilde{A}_{10} = 1000\, e^{(0.06)(10)} = 1000\, e^{0.6} = 1822.12 \tag{5.67}$$

where we have used the approximate value 2.71828 for e (whose derivation will be given later) and the value of $\widetilde{A}_{10}$ shows two decimal figures. The amount $\widetilde{A}_{10}$ is seen to exceed the amounts $A_{10,1} = 1790.85$, $A_{10,2} = 1806.11$, and $A_{10,4} = 1814.02$. Indeed, it is the maximum amount 1000 dollars can generate in 10 years at 6 percent per annum. In this connection see Exercise 5.26.

If 100 dollars were left in the bank for only one year, the continuous compounding of interest at 6 percent per annum would result in

$$\widetilde{A}_1 = 100\, e^{0.06} = 106.18 \tag{5.68}$$

Equation (5.68) shows that 6 percent per annum compounded continuously is effectively equivalent to 6.18 percent.

As a final example, we reflect that one dollar deposited for one year at 100 percent annual interest would become two dollars at the end of the year. However, if the 100 percent interest was compounded continuously, the single dollar obeying (5.66) would become e dollars, that is, approximately 2.72 dollars. It is quite amazing what continuous compounding of interest can do to a dollar in the course of one year!

5.9 Exponential decay

The discussion in the previous section applies to many natural phenomena associated with the growth described by (5.66), known as *exponential growth.* We shall see now that the number e enters the picture in processes of depreciation or decay as well. Suppose, for example, that the income tax authorities allow a deduction from the income of a businessperson of twenty percent per annum of the value of the car used for work, due to depreciation of the car. Accordingly, if the car was bought for A_0 dollars, its value after n years – as reckoned by the income tax authorities – would be

$$A_{n,1} = A_0 \left(1 - \frac{20}{100}\right)^n \tag{5.69}$$

where $A_{n,1}$ has a meaning similar to $A_{10,1}$ in (5.42), except that here we have a minus sign because the value decreases rather than increases. A quick calculation based upon (5.69) shows that at twenty percent depreciation per year, the car would reach half its original value after approximately 3 years. Specifically, putting $A_{n,1} = \frac{1}{2}A_0$, we obtain

$$A_{n,1} = A_0 \left(1 - \frac{20}{100}\right)^n = \frac{1}{2}A_0 \tag{5.70}$$

or

$$(0.8)^n = 0.5 \tag{5.71}$$

Taking the logarithms of both sides of (5.71), we find n to be approximately equal to 3.1. More generally, if the depreciation were reckoned k times a year at p percent annually, we would have, similar to (5.64),

$$A_{n,k} = A_0 \left(1 - \frac{\alpha}{k}\right)^{kn} \tag{5.72}$$

with $\alpha = p/100$, α being 0.2 in our specific case. Moreover, if depreciation were to take place continuously, considerations similar to those leading to (5.66) yield

$$\widetilde{A}_n = A_0 e^{-\alpha n} \tag{5.73}$$

This is the formula expressing *exponential decay*, equivalent to continuous depreciation. Under exponential decay our car would reach half its original value after about 3.5 years, because

$$\widetilde{A}_n = A_0 e^{-(0.2)n} = \frac{1}{2} A_0 \tag{5.74}$$

implies

$$e^{-(0.2)n} = 0.5 \tag{5.75}$$

so n approximately equals 3.5. Note that exponential decay is slower than the decay associated with annual depreciation (for equal annual percentage) in contrast with exponential growth, which is faster than the growth associated with annual compounding of interest. The reason for this becomes clear if we reflect that an annual depreciation of twenty percent leaves the value of the car at $(0.80)A_0$ by the end of one year. However, depreciation of ten percent every half-year leaves the car with a value of $(0.90)A_0$ after the first six months; and because ten percent of ninety percent is nine percent, the car is worth $(0.81)A_0$ at the end of the year. This effect is even more pronounced if depreciation takes place quarterly. Eventually, at continuous depreciation (see also Exercise 5.27), the value of the car at the end of one year is

$$A_0 e^{-0.2} = (0.8187)A_0 \tag{5.76}$$

It follows that continuous depreciation is indeed slower and effectively equals an annual depreciation of 18.13 percent, and not 20 percent, because

$$(0.8187)A_0 = A_0 \left(1 - \frac{18.13}{100}\right) \tag{5.77}$$

The rate of exponential decay can be assessed in several ways. One way is to ask for the time (the value of n) elapsed before the quantity under consideration decays to half its original value. This time, the *half-life*, plays a useful role in various disciplines. To find the half-life, we set $\widetilde{A}_n = A_0/2$ in (5.73) and obtain

$$A_0 e^{-\alpha n} = \frac{1}{2} A_0 \tag{5.78}$$

Note that the half-life n in (5.78) is independent of A_0, which cancels out. Thus,

$$e^{\alpha n} = 2 \tag{5.79}$$

from which

$$\alpha n = \log_e 2 \tag{5.80}$$

The notation $\log_e 2$ denotes the "natural logarithm" of 2, that is, the logarithm to the base e (rather than 10). We see that the natural logarithm has arisen quite naturally and, as a matter of fact, appears in many other applications as well. On account of its frequent occurrence, it is denoted by the special symbol "ln", that is, $\log_e x = \ln x$. Accordingly, (5.80) states that the half-life n is given by

$$n = \frac{\ln 2}{\alpha} = \frac{0.693147}{\alpha} \tag{5.81}$$

where the value of ln 2 has been obtained by the area computation methods of Chapter 3, such as in Exercise 3.2.

Equation (5.73), as already pointed out, has many applications in areas other than yearly growth or depreciation. A radioactive material, for example, is known to undergo decay satisfying the law

$$A(t) = A_0 e^{-\alpha t} \tag{5.82}$$

where t stands for time and $A(t)$ is the quantity to which A_0 has decayed by time t. Moreover, α now stands for a material constant. (Note that $A(0) = A_0$.) The half-life, usually denoted by $t_{1/2}$, is given by $t_{1/2} = (1/\alpha)\ln 2$ as before. Radioactive materials that have half-lives of the order of thousands of years are used in archaeology to help determine the age of excavated finds. Medical science uses half-lives of the order of hours to help in scanning of various organs of the human body. Finally, materials science, nuclear physics, and chemistry often use radioactive isotopes that have half-lives of a few microseconds.

As is evident, the number e merits careful study. We saw earlier that $2 < e < 3$, but we now want to show how to compute e efficiently to any desired degree of accuracy, which will justify our use of the value $e \approx 2.71828$.

5.10 The computation of e

With a view toward the efficient computation of e, let us recall that $a_k = (1+1/k)^k$, as defined in (5.84), has been shown to be an increasing bounded sequence that tends to e. Moreover, if we set

$$b_k = 1 + \frac{1}{1!} + \frac{1}{2!} + \cdots + \frac{1}{k!} = \sum_{m=0}^{k} \frac{1}{m!} \tag{5.83}$$

it follows from (5.55)–(5.60) that

$$a_k \leq b_k < 3 \tag{5.84}$$

Obviously, b_k is an increasing sequence since $b_k - b_{k-1} = 1/k! > 0$. Because b_k is increasing and bounded, it tends to a limiting value that will be denoted by e^* (see also Exercise 5.28). We shall presently show that $e^* = e$, thus deriving an alternative representation for e, that is,

$$1 + \frac{1}{1!} + \frac{1}{2!} + \cdots + \frac{1}{k!} \longrightarrow e^* = e \tag{5.85}$$

Moreover, it turns out that the sequence b_k lends itself more easily than the sequence a_k to the computation of e, including error estimate and control (which is a difficult task for the sequence a_k).

To derive the result $e^* = e$, observe first that because the inequality (5.84) is valid for all k, it follows (from the definition of the limit concept, studied in calculus) that

$$e \leq e^* \tag{5.86}$$

The desired result $e = e^*$ will have been shown, once we show that $e \geq e^*$ as well. To see this we observe, on the basis of (5.53), that for every integer r satisfying $r < k$, we have

$$a_k \geq 1 + 1 + \frac{1}{2!}\left(1 - \frac{1}{k}\right) + \cdots + \frac{1}{r!}\left(1 - \frac{1}{k}\right)\left(1 - \frac{2}{k}\right)\cdots\left(1 - \frac{r-1}{k}\right) \tag{5.87}$$

because the $(k - r)$ neglected terms are all nonnegative. Letting k tend to infinity in (5.87) while holding r fixed for the time being, we reach

$$e \geq 1 + 1 + \frac{1}{2!} + \cdots + \frac{1}{r!} = b_r \tag{5.88}$$

Finally, we let r tend to infinity in (5.88) and obtain

$$e \geq e^* \tag{5.89}$$

This and (5.86) imply that $e = e^*$, as desired.

Having proved that the number e is the limiting value of the sequence b_k, we can write

$$\begin{aligned} e &= b_k + \left[\frac{1}{(k+1)!} + \frac{1}{(k+2)!} + \cdots\right] \\ &= b_k + t_k \end{aligned} \tag{5.90}$$

with t_k denoting the expression in the square brackets. In other words, t_k is the error (in this case, the tail of the series) incurred when we approximate e by b_k. We may estimate t_k by using a majorizing geometric series, in a manner analogous to (5.55)–(5.60), to obtain

$$\begin{aligned} t_k &= \frac{1}{(k+1)!} + \frac{1}{(k+2)!} + \frac{1}{(k+3)!} + \cdots \\ &= \frac{1}{(k+1)!}\left[1 + \frac{1}{k+2} + \frac{1}{(k+2)(k+3)} + \cdots\right] \\ &< \frac{1}{(k+1)!}\left[1 + \frac{1}{k+1} + \frac{1}{(k+1)^2} + \cdots\right] \\ &= \frac{1}{(k+1)!}\left[\frac{1}{1 - 1/(k+1)}\right] = \frac{\frac{1}{k}}{k!} \end{aligned} \tag{5.91}$$

Thus, we reach

$$0 < e - b_k < \frac{\frac{1}{k}}{k!} \tag{5.92}$$

and we are in the position to control the size of the error by choosing k sufficiently large. For example, suppose we want to compute e correct to six decimal places. Then we must have $(1/k)/k! < (1/2) \cdot 10^{-6}$ or $2 \cdot 10^6 < k \cdot k!$, which is achieved already for $k = 9$. In this connection see Exercises 5.29 and 5.30. An algorithm for the computation of e to any desired accuracy can take the following form:

1. Input q (the number of desired correct decimal figures).
2. Set $B = 1$, $T = 1$, and $k = 0$.
3. Increase k by 1.
4. Replace the value of T by T/k.
5. Replace the value of B by $B + T$.
6. Print the value of k and B.
7. If $(1/k)T > (1/2) \cdot 10^{-q}$, then return to Step 3.
8. End.

The laboratory participants should note that in Step 4 the variable T is computed to equal $1/k!$ by dividing its previously computed value $1/(k-1)!$ by k. Thus, in Step 5 the variable B accumulates the sum defining b_k. Moreover, the quantity $(1/k)T$ in Step 7 represents the error bound (5.92). We ran a program based on this algorithm on a PC with double precision and obtained the results shown in Table 5.7, for $q = 15$. The last value, obtained for $k = 17$, is indeed correct to 15 decimal figures, illustrating the efficiency of the computation of e by means of the sum b_k in (5.83). In this connection see Exercise 5.31.

Table 5.7. *Approximations of e*

k	$B = b_k$
1	2.000000000000000
2	2.500000000000000
3	2.666666666666667
4	2.708333333333333
5	2.716666666666667
6	2.718055555555556
7	2.718253968253968
8	2.718278769841270
9	2.718281525573192
10	2.718281801146385
11	2.718281826198493
12	2.718281828286169
13	2.718281828446759
14	2.718281828458230
15	2.718281828458995
16	2.718281828459042
17	2.718281828459045

We have seen in this chapter how to calculate π and e. The laboratory participants will encounter these numbers throughout their mathematical and scientific activities. The computational laboratory assignments of this chapter should hence be part and parcel of their fundamental mathematical education.

Exercises

5.1 Compute the circumferences of the regular hexagons that are inscribed in and circumscribed about the unit circle, and hence show that $3 < \pi < 2\sqrt{3}$, as in (5.1).

5.2 Express the length of the side of the regular n-gon circumscribing the unit circle in terms of the length of the side of the corresponding circumscribed n-gon, and thus arrive at (5.5) and (5.6). Justify each of your steps carefully with geometrical arguments.

5.3 Employ geometrical arguments to show that the sequence $3a_6$, $6a_{12}, 12a_{24}, \ldots$ is increasing and bounded and the sequence $3c_6$, $6c_{12}, 12c_{24}, \ldots$ is decreasing and bounded.

5.4 Show that for values of u satisfying $0 \leq u \leq 1$, we have $1/\sqrt{1-u/4} \leq 1 + u/6$. After you have shown this inequality, replace the number 6 by the parameter α and find the largest α such that the inequality still holds for $0 \leq u \leq 1$.

5.5 Construct an Archimedean process, starting with squares rather than hexagons. Here the initial sides are $a_4 = \sqrt{2}$ and $c_4 = 2$; hence you must modify inequalities (5.9)–(5.14) appropriately.

5.6 Modify the Archimedean algorithm for the approximation of π so that it starts with squares rather than hexagons. Write and run a corresponding program and compare the numerical results with those in Table 5.1.

5.7 Run the Archimedean algorithm in which the condition $E \geq t$ in Step 8 has been replaced by the condition $M \geq n$ (for suitably large M) and see for yourself how contamination sets in and develops. For an even better demonstration, have the program print out the difference between each approximation obtained and a very accurate value of π.

5.8 Prove that for values of h satisfying $0 \leq h \leq \frac{1}{2}$, we have

$$\frac{2(1+h)}{\sqrt{3}+\sqrt{3-4h-4h^2}} > \frac{h}{\sqrt{1+h^2}+1}$$

Interpret this inequality geometrically in connection with the behavior of $y = \sqrt{1-x^2}$ near the endpoints of the interval $[0, \frac{1}{2}]$.

5.9 Write and run a program for the computation of π by means of the trapezoidal method applied to $f(x) = \sqrt{1-x^2}$, $0 \leq x \leq \frac{1}{2}$, ensuring $q = 4,\ 5$, and 6 correct decimal figures.

5.10 Suppose we roll a fair die. What is the probability of getting (a) the number 3, (b) an even number, (c) either the number 1 or the number 6, (d) the number 8, (e) a whole number less than 7, (f) anything but 4. Justify your answers.

5.11 Now, suppose we roll two fair dice. What is the probability of getting (a) the number 4, (b) the number 7, (c) a number larger than 8. Justify your answers. What would your answers to (a),(b),(c) have been, had you rolled three dice.

5.12 Suppose that, in addition to $RAND(x)$, the computer's library contains the built-in function $INT(x)$, giving the greatest integer not exceeding x. Explain why we may use the function $INT(RAND(6)+1)$ to simulate the rolling of a die. Write and run a program that actually simulates 6000 rollings of a die and prints out the relative frequencies of the various outcomes.

5.13 Repeat Exercise 5.12 for the simulation of 6000 rollings of two dice.

5.14 Write and run a program that computes an approximation for one quarter of the unit circle's area in the following manner:

Generate pairs of random numbers x and y such that $0 \le x < 1$, $0 \le y < 1$, and think of each pair (x, y) as coordinates of a point in this unit square. Now count the number of cases in which the point (x, y) is situated inside the quarter circle. In this way (known as the Monte-Carlo method) it is possible to approximate the area under the graph of any nonnegative function $f(x)$.

5.15 Exercise 5.14 actually simulates rain falling randomly into one container with a circular cross section and into another container with a square cross section (whose side equals the diameter of the circle). Find an approximation for π by actually placing two such containers in the rain and weighing the respective quantities of water after a rainy night. Repeat the experiment, using successively longer periods of rainfall.

5.16 Carry out, manually, a Buffon's needle experiment with T tosses for $T = 1000, 2000, 3000$. You can use, for example, 50 toothpicks of uniform length L for each toss, and have the toothpicks land on any striped floor (with uniform stripe width W). Recall that in this case the estimate of π equals $(2T/C) \cdot (L/W)$ and explain why.

5.17 Write and run a program simulating the Buffon experiment up to a million tosses and print out results every 10,000 tosses.

5.18 Using (5.32), show that

$$\frac{\pi}{4} = 2 \arctan \frac{1}{5} + \arctan \frac{1}{7} + 2 \arctan \frac{1}{8}$$

as well as that

$$\frac{\pi}{4} = 5 \arctan \frac{1}{8} + 2 \arctan \frac{1}{18} + 3 \arctan \frac{1}{57}$$

[Hint: set $u = 1/m$ and $v = 1/n$ in (5.32).]

5.19 Suppose we compute π using (5.36) and (5.37). Determine the number k of terms required to ensure q correct decimal figures.

5.20 Write and run a program for the computation of π by means of (5.36) and (5.37). Verify for $q = 4$, 5, and 6 that the number k of terms you determined in Exercise 5.19 actually gives the desired accuracy

5.21 Repeat Exercise 5.20 using your own homemade formula instead of (5.36).

5.22 Use the definition (5.50) of the binomial coefficients to prove that

$$\binom{m}{r} = \binom{m}{m-r} = \frac{m(m-1)(m-2)\cdots(m-r+1)}{r!}$$

$$\binom{m}{0} = \binom{m}{m} = 1$$

$$\binom{m}{1} = \binom{m}{m-1} = m$$

5.23 Using (5.53), show that $a_{k+1} > a_k \geq 2$.

5.24 Recall the arithmetic–geometric inequality

$$\sqrt[m]{t_1 \cdot t_2 \cdot t_3 \cdot \cdots \cdot t_m} \leq \frac{t_1 + t_2 + t_3 + \cdots + t_m}{m}$$

which for $m = 2$ follows directly from $\left(\sqrt{t_1/2} - \sqrt{t_2/2}\right)^2 \geq 0$ and can be generalized by mathematical induction. Now choose $m = k+1$, $t_1 = 1$, $t_2 = t_3 = \cdots = t_{k+1} = 1 + 1/k$ and hence show that the sequence $a_k = (1 + 1/k)^k$ is increasing. (This exercise and the next one have been suggested to us by Dr. B. Arbel.)

5.25 In the arithmetic–geometric inequality, choose $m = k+2$, $t_1 = t_2 = \frac{1}{2}$, $t_3 = t_4 = \cdots = t_{k+2} = 1 + 1/k$ and thus show that the sequence $a_k = (1+1/k)^k$ is bounded by 4. Repeat this process with $m = k+6$, $t_1 = t_2 = \cdots = t_6 = \frac{5}{6}$, $t_7 = t_8 = \cdots = t_{k+6} = 1 + 1/k$ and obtain a sharper upper bound for a_k. Sharpen this bound once more, using the fraction $\frac{10}{11}$.

5.26 How many years must elapse until an initial amount A_0 deposited at 10 percent per annum will double itself if interest is compounded (a) annually, (b) quarterly, (c) monthly, and (d) continuously. What is the effective annual yield corresponding to 10 percent compounded continuously. Use the approximate value 2.71828 for e.

5.27 How many years must elapse until a piece of equipment, depreciating 10 percent annually, will reach a quarter of its original value if depreciation is reckoned (a) annually, (b) quarterly, (c) monthly, and (d) continuously. What is the effective annual loss corresponding to 10 percent continuous depreciation.

5.28 Given the sequence

$$c_1 = 1, \qquad c_2 = 1 + \frac{1}{2^2}, \qquad \ldots, \qquad c_n = \sum_{k=1}^{n} \frac{1}{k^2}$$

Show that this sequence is increasing and bounded and hence tends to a limiting value c, where $1 < c < 2$.
[Hint: $1/k^2 < 1/[k(k-1)] = 1/(k-1) - 1/k$, for all $k \geq 2$.]

5.29 If e is approximated by $A = 2 + \frac{1}{2} + \frac{1}{6} + \frac{1}{24} + \frac{1}{120} = \frac{163}{60}$, estimate the error $|e - A|$.

5.30 For which value of k is the error $|e - b_k|$ smaller than (a) $\frac{1}{2} \cdot 10^{-4}$, (b) $\frac{1}{2} \cdot 10^{-8}$, (c) $\frac{1}{2} \cdot 10^{-12}$.

5.31 Write and run your own program for computing e through the use of the sequence b_k.

6
Convergence acceleration

6.1 Introduction

The solution of many mathematical problems requires the summation of infinite series. In problems that involve infinite geometric series the formula for the sum is known, but in many other cases, no closed-form formula for the sum is available. Moreover, in many cases the rate of convergence is too slow for the practical application of the series. Consider an example in which the solution requires the summation of a slowly converging infinite series: On a bright Saturday morning Johnny walked for one hour, covering a distance d with constant velocity v. On Sunday, he walked half that distance with velocity $\frac{1}{2}v$, he then walked one third of the remaining distance at velocity $\frac{2}{3}v$, one quarter of the remaining distance with velocity $\frac{3}{4}v$, and so on. We shall show that on Sunday Johnny actually covered the same distance as on Saturady, and shall find how long it took. Clearly, without loss of generality, we can assume that $v = 1$ and $d = 1$, in appropriate units. Thus, for the Sunday walk, we obtain Table 6.1.

The total distance covered on Sunday is thus given by

$$\frac{1}{1 \cdot 2} + \frac{1}{2 \cdot 3} + \frac{1}{3 \cdot 4} + \cdots + \frac{1}{n(n+1)} + \cdots = \sum_{n=1}^{\infty} \frac{1}{n(n+1)} \tag{6.1}$$

This is an example of an infinite series that can be summed exactly, and its sum will later be shown to equal 1, thus proving that the total Sunday distance equals the Saturday distance. Now, let us denote by t_n the duration of Sunday's nth part of the walk. Thus,

$$t_n = \frac{1/[n(n+1)]}{n/(n+1)} = \frac{1}{n^2} \tag{6.2}$$

Table 6.1. *Sunday's walk*

#	Velocity	Distance	The remaining distance
1	$\frac{1}{2}$	$\frac{1}{2}$	$1-\frac{1}{2}=\frac{1}{2}$
2	$\frac{2}{3}$	$\frac{1}{2}\cdot\frac{1}{3}$	$\frac{1}{2}-\frac{1}{6}=\frac{1}{3}$
3	$\frac{3}{4}$	$\frac{1}{3}\cdot\frac{1}{4}$	$\frac{1}{3}-\frac{1}{12}=\frac{1}{4}$
$\vdots$	$\vdots$	$\vdots$	$\vdots$
n	$\frac{n}{n+1}$	$\frac{1}{n}\cdot\frac{1}{n+1}$	$\frac{1}{n+1}$
$\vdots$	$\vdots$	$\vdots$	$\vdots$

so the total time spent on Sunday's walk equals

$$t = \sum_{n=1}^{\infty} \frac{1}{n^2} \tag{6.3}$$

To compute t to a desirable accuracy, a very large and impractical number of terms must be used. This particular series will be assigned to the laboratory participants for the purpose of convergence acceleration, as outlined in the following sections.

The slow convergence of many series motivates a search for methods that accelerate the convergence process. We shall present various acceleration methods here, using only rudimentary knowledge of infinite series, the central purpose of each method being to reduce the number of terms needed to achieve a desired accuracy.

The sum of the first k terms of our infinite series is denoted, as is customary in calculus, by

$$S_k = \sum_{n=1}^{k} a_n \tag{6.4}$$

The numbers $S_1, S_2, S_3, \ldots$ form a sequence known as the sequence of partial sums. If this sequence approaches a limiting value S, we employ the usual notation

$$S = \sum_{n=1}^{\infty} a_n \tag{6.5}$$

and say that the series converges to the sum S. We also use the notation

$$R_k = S - S_k = \sum_{n=k+1}^{\infty} a_n \tag{6.6}$$

for the remainder, that is, the difference between the sum of the series and the kth partial sum.

In practice, if we want to find the sum of a convergent series to a prescribed accuracy, we are mainly interested in how many terms are needed to ensure this accuracy. A method that enables us to significantly reduce the number of required terms is called an *acceleration method.* Before we study acceleration methods, we must clarify and illustrate the concept of *rate of convergence*.

6.2 Rate of convergence

To get a feeling for the rate of convergence, we start with a convergent series whose terms alternate in sign and decrease in absolute value:

$$S = \sum_{n=1}^{\infty} (-1)^n a_n \tag{6.7}$$

with

$$a_n > 0, \qquad a_{n+1} < a_n \tag{6.8}$$

The remainder R_k, after the first k terms, satisfies

$$\begin{aligned} |R_k| &= (a_{k+1} - a_{k+2}) + (a_{k+3} - a_{k+4}) + \cdots \\ &= a_{k+1} - (a_{k+2} - a_{k+3}) - (a_{k+4} - a_{k+5}) - \cdots \end{aligned} \tag{6.9}$$

so that by (6.8) we have

$$a_{k+1} - a_{k+2} < |R_k| < a_{k+1} \tag{6.10}$$

In particular, the remainder of such a series is bounded by the absolute value of the first truncated term. Because the series was assumed to be convergent, we know that $a_{k+1} \to 0$ with increasing k, and hence the remainder R_k can be made as small as we please. The quantity a_{k+1} serves here as a natural bound for the error incurred when the sum S is approximated by the partial sum S_k. Clearly, an error can only be bounded, not computed exactly because otherwise it could be added to the calculated result and would no longer be an error.

We are now in a position to compare the rates of convergence of two infinite series. For example, let us choose

$$S = \sum_{n=1}^{\infty} \frac{(-1)^n}{n}, \qquad \tilde{S} = \sum_{n=1}^{\infty} \frac{(-1)^n}{n^2} \tag{6.11}$$

and find S and $\tilde{S}$ with a tolerance of 10^{-6}, that is, with errors less than 10^{-6}. To ensure this tolerance for the first series, we must have

$$|R_k| < a_{k+1} = \frac{1}{k+1} < 10^{-6} \tag{6.12}$$

which means that a million terms are needed. This, of course, is unacceptable from a practical point of view. For the second series, we must require

$$|\tilde{R}_k| < \tilde{a}_{k+1} = \frac{1}{(k+1)^2} < 10^{-6} \tag{6.13}$$

so that a thousand terms suffices. Thus, we say that the second series converges faster than the first.

To express this fact formally, it is customary to use the symbols O and o. Assume that $\tilde{R}_k$ and R_k tend to zero as k tends to infinity. The notation $\tilde{R}_k = O(R_k)$ means that $(\tilde{R}_k/R_k) \to c \neq 0$, expressing the fact that $\tilde{R}_k$ and R_k approach zero at the same rate, and hence they are said to be of the same order of magnitude. If $c = 0$, we say that $\tilde{R}_k$ is of a smaller order of magnitude than R_k (i.e., it tends to zero faster), and we use the notation $\tilde{R}_k = o(R_k)$ as k tends to infinity. More general definitions of the symbols O and o could be given, but this definition satisfies our purposes. At this point it is recommended that the laboratory participants familiarize themselves with the use of these symbols through a set of examples. In this connection see Exercises 6.1 and 6.2.

When we compare the rate of convergence of two convergent series $\sum a_n$ and $\sum b_n$, we are actually comparing the behavior of their remainders. The fact that the series converge at the same rate is expressed by

$$\sum_{n=k+1}^{\infty} b_n = O\left(\sum_{n=k+1}^{\infty} a_n\right) \tag{6.14}$$

whereas the fact that the series $\sum b_n$ converges faster than $\sum a_n$ is expressed by

$$\sum_{n=k+1}^{\infty} b_n = o\left(\sum_{n=k+1}^{\infty} a_n\right) \tag{6.15}$$

For obvious reasons, if we want to compute the sum of a series $\sum a_n$, we are interested in finding another series $\sum b_n$ that converges to the same value but more rapidly. This is the purpose of convergence-acceleration methods, of which some typical examples are presented in the following sections.

6.3 Closed-form sums

Series whose sum can be expressed in closed form turn out to be useful in the evaluation of the sum of other series, as will be shown later. Such closed-form series are, for example, decreasing geometric series (see Exercise 6.3).

The following is a way of generating an infinite series that converges to an arbitrary sum S. Set $S_n = S - x_n$, where x_n is any sequence tending to zero. Then the series

$$S_1 + (S_2 - S_1) + (S_3 - S_2) + \cdots + (S_n - S_{n-1}) + \cdots \tag{6.16}$$

converges to S. For example, for $S = 1$ and $x_n = 1/(n+1)$, we obtain the series

$$\begin{aligned} &\frac{1}{2} + \left(\frac{2}{3} - \frac{1}{2}\right) + \cdots + \left(\frac{n}{n+1} - \frac{n-1}{n}\right) + \cdots \\ &\qquad = \frac{1}{1\cdot 2} + \frac{1}{2\cdot 3} + \frac{1}{3\cdot 4} + \cdots = \sum_{n=1}^{\infty} \frac{1}{n(n+1)} \end{aligned} \tag{6.17}$$

The fact that this series actually converges to $S = 1$ can be seen directly as follows. We observe that $1/[n(n+1)] = 1/n - 1/(n+1)$, and thus we can write

$$\begin{aligned} S_k &= \sum_{n=1}^{k} \frac{1}{n(n+1)} = \sum_{n=1}^{k} \left(\frac{1}{n} - \frac{1}{n+1}\right) \\ &= 1 - \frac{1}{2} + \frac{1}{2} - \frac{1}{3} + \frac{1}{3} - \frac{1}{4} + \cdots + \frac{1}{k} - \frac{1}{k+1} \\ &= 1 - \frac{1}{k+1} \end{aligned} \tag{6.18}$$

Letting k tend to infinity, we obtain the desired result $S_k \to S = 1$. A series such as (6.18), whose terms cancel in pairs except the first one – which turns out to be the value of the sum – is referred to as a *telescoping series*. The laboratory participants should familiarize themselves with series that converge to a given sum, leading to telescoping series (see Exercises 6.4 and 6.5).

6.4 Upper and lower remainder estimates

In many cases, upper and lower estimates of the remainder of a given series can be found. When this is so, these estimates can be used to make the summation of the series more efficient; that is, they enable us to compute the sum to a given accuracy with fewer terms. We show this process by considering $S = \sum(1/n^2)$, the series arose in our example of Sunday's walk. In this case, the remainder can be bounded from above as follows:

$$\begin{aligned} R_k = \sum_{n=k+1}^{\infty} \frac{1}{n^2} &= \frac{1}{(k+1)^2} + \frac{1}{(k+2)^2} + \cdots \\ &< \frac{1}{k(k+1)} + \frac{1}{(k+1)(k+2)} + \cdots \\ &= \frac{1}{k} - \frac{1}{k+1} + \frac{1}{k+1} - \frac{1}{k+2} + \cdots \end{aligned} \tag{6.19}$$

This last series is telescoping, and its sum equals $1/k$. Thus,

$$R_k < 1/k \tag{6.20}$$

To find S and be sure that the error is less than $\frac{1}{2} \cdot 10^{-4}$ (i.e., assure four correct decimal digits), 20,000 terms are needed. This is impractical. Fortunately, we can also find a lower bound for the remainder because

$$\begin{aligned} R_k &> \frac{1}{(k+1)(k+2)} + \frac{1}{(k+2)(k+3)} + \cdots \\ &= \frac{1}{k+1} - \frac{1}{k+2} + \frac{1}{k+2} - \frac{1}{k+3} + \cdots = \frac{1}{k+1} \end{aligned} \tag{6.21}$$

From (6.20) and (6.21) we obtain

$$\frac{1}{k+1} < R_k < \frac{1}{k} \tag{6.22}$$

and, using the average of these bounds, we can write

$$R_k = \frac{1}{2}\left(\frac{1}{k} + \frac{1}{k+1}\right) + E = \frac{k + \frac{1}{2}}{k(k+1)} + E \tag{6.23}$$

where

$$|E| < \frac{1}{2}\left(\frac{1}{k} - \frac{1}{k+1}\right) = \frac{\frac{1}{2}}{k(k+1)} < \frac{1}{2k^2} \tag{6.24}$$

The laboratory instructor should point out the following general principle: Whenever some desired quantity can be bounded from above and from below, we can approximate it by the average of the bounds, knowing that the error in absolute value does not exceed half their difference. Moreover, in this particular example, the upper and lower bounds of the remainder are so close that their difference is proportional to $1/k^2$, whereas the bounds themselves are merely proportional to $1/k$. Thus, using this new error estimate, we now have

$$S = \sum_{n=1}^{k} \frac{1}{n^2} + \frac{k + \frac{1}{2}}{k(k+1)} + E \tag{6.25}$$

where $|E| < 1/(2k^2)$. Making use of (6.25) rather than (6.20), we need only 100 terms to make sure that the error is less than $\frac{1}{2} \cdot 10^{-4}$. Thus, using the upper and lower bounds of the remainder, we are able to reduce the number of terms from 20,000 to 100 for a tolerance of $\frac{1}{2} \cdot 10^{-4}$. To reduce the number of terms even further, we develop other acceleration techniques, introduced in the following sections.

To practice using upper and lower estimates of the remainder, the laboratory participants should analyze other series, such as $\sum(1/n^3)$. The final result in this case turns out to be

$$\sum_{n=1}^{\infty} \frac{1}{n^3} = \sum_{n=1}^{k} \frac{1}{n^3} + \frac{k^2 + k + 1}{2(k-1)k(k+1)(k+2)} + E \tag{6.26}$$

where $|E| < 1/(k-1)^3$. On the other hand, using only the upper estimate of the remainder, we obtain an error bound of $1/[2(k-1)^2]$. In this connection see Exercises 6.6 and 6.7.

Bounding the remainder from both sides is not always possible. When it is, though, summation efficiency can be considerably improved, particularly when the relevant series converges slowly and the bounds are sufficiently close.

We emphasize that the intrinsic round-off errors of the microcomputers used in the laboratory must also be taken into account when the summation of series is actually carried out. This will be demonstrated presently, once again using the infinite series $\sum(1/n^2)$. Thus, to find $S = \sum(1/n^2)$ to four correct decimal figures, we used (6.25) with $k = 100$ on a microcomputer having single precision of seven figures. Running a short program, we obtained $S = 1.6449$ within a few seconds. (The laboratory instructor, knowing from Fourier series that $\sum(1/n^2) = \pi^2/6$, will find 1.6449 to be correct to that accuracy.) In juxtaposition, we summed the first 20,000 terms of $\sum(1/n^2)$ as is, without any attempt at efficiency. Not only did the computation take about half an hour, but it yielded $S = 1.6447$ to four decimal digits, falsifying the last digit. This falsification, of course, was caused by round-off errors, stemming from the fact that after a few thousand terms, additional terms no longer affected the computed result. This can be demonstrated in the laboratory by printing out results, say, every 200 terms. After approximately 5000 terms the sum reached 1.644725 and did not change thereafter (as noted, the microcomputer we used had only seven figures). When we repeated the computation with double precision, however, we obtained $S = 1.644884$ (i.e., $S = 1.6449$ to four decimal figures), and computation time was almost an hour (see Excercise 6.8). The search for acceleration methods is thus well motivated not only by efficiency considerations but also by the necessity of avoiding round-off inaccuracies.

6.5 Kummer's method

As mentioned earlier, the final goal of any acceleration method is the replacement of the original, slowly converging series by another series that converges faster (to the same sum).

We begin by studying the method suggested by Ernst Kummer (1810–1893), in which we subtract from the series $\sum a_n$ another series $\sum b_n$ that has "similar behavior" (in a sense to be explained presently) and whose sum is expressible in closed form. At that point we still have to sum an infinite series that equals the difference between the original and the similar series, but which turns out to converge faster than $\sum a_n$. By *similar behavior* we mean that $(a_n/b_n) \to 1$, as is the case with $\sum(1/n^2)$ and $\sum 1/[n(n+1)]$. Because, as we have seen before,

$$\sum_{n=1}^{\infty} \frac{1}{n(n+1)} = 1 \tag{6.27}$$

we can write

$$\sum_{n=1}^{\infty}\frac{1}{n^2}=\sum_{n=1}^{\infty}\frac{1}{n(n+1)}+\sum_{n=1}^{\infty}\left[\frac{1}{n^2}-\frac{1}{n(n+1)}\right]$$
$$=1+\sum_{n=1}^{\infty}\frac{1}{n^2(n+1)} \tag{6.28}$$

The latter series is the one that remains to be summed, and indeed, it converges faster than $\sum(1/n^2)$. Thus, its remainder $\tilde{R}_k$ can be estimated by a telescoping series, so that

$$\tilde{R}_k=\sum_{n=k+1}^{\infty}\frac{1}{n^2(n+1)}\leq\sum_{n=k+1}^{\infty}\frac{1}{(n-1)n(n+1)}$$
$$=\sum_{n=k+1}^{\infty}\left[\frac{1/2}{(n-1)n}-\frac{1/2}{n(n+1)}\right]=\frac{1/2}{k(k+1)}<\frac{1}{2k^2} \tag{6.29}$$

Collecting our results, we find that the new remainder is bounded by $1/(2k^2)$ compared with $1/k$, which by (6.20) bounds the remainder of $\sum(1/n^2)$.

The laboratory participants should now realize that it is possible to go a step further and apply Kummer's method to the new series as well. This can evidently be done by subtracting from it the series $\sum 1/[n(n+1)(n+2)]$, whose sum can be found at once. Thus,

$$\sum_{n=1}^{\infty}\frac{1}{n(n+1)(n+2)}=\sum_{n=1}^{\infty}\left[\frac{1/2}{n(n+1)}-\frac{1/2}{(n+1)(n+2)}\right]=\frac{1}{4} \tag{6.30}$$

Consequently,

$$\sum_{n=1}^{\infty}\frac{1}{n^2(n+1)}=\frac{1}{4}+\sum_{n=1}^{\infty}\left[\frac{1}{n^2(n+1)}-\frac{1}{n(n+1)(n+2)}\right]$$
$$=\frac{1}{4}+\sum_{n=1}^{\infty}\frac{2}{n^2(n+1)(n+2)} \tag{6.31}$$

and so, by (6.28) and (6.31), we have

$$\sum_{n=1}^{\infty}\frac{1}{n^2}=\frac{5}{4}+\sum_{n=1}^{\infty}\frac{2}{n^2(n+1)(n+2)} \tag{6.32}$$

We could apply Kummer's method several more times, but the acceleration achieved so far is already considerable. This can be seen from the remainder estimate.

$$\begin{aligned}\tilde{\tilde{R}}_k &= \sum_{n=k+1}^{\infty} \frac{2}{n^2(n+1)(n+2)} \le \sum_{n=k+1}^{\infty} \frac{2}{(n-1)n(n+1)(n+2)} \\ &= \sum_{n=k+1}^{\infty} \left[\frac{2/3}{(n-1)n(n+1)} - \frac{2/3}{n(n+1)(n+2)}\right] \\ &= \frac{2/3}{k(k+1)(k+2)} < \frac{2/3}{k^3} \qquad (6.33)\end{aligned}$$

At this point, the laboratory participants should compare the number of terms needed to assure a given accuracy, say $\frac{1}{2} \cdot 10^{-4}$, for each of the three relevant series. This demonstrates the practical consequences of the acceleration process.

For $\sum \frac{1}{n^2}$ *we require* $\frac{1}{k} < \frac{1}{2} \cdot 10^{-4}$, *so* 20,000 *terms are needed.*

For $1 + \sum \frac{1}{n^2(n+1)}$ *we require* $\frac{1}{2k^2} < \frac{1}{2} \cdot 10^{-4}$, *so* 100 *terms are needed.*

For $\frac{5}{4} + \sum \frac{1}{n^2(n+1)(n+2)}$ *we require* $\frac{2/3}{k^3} < \frac{1}{2} \cdot 10^{-4}$, *so* 24 *terms are needed.*

Using 24 terms of the third series on a microcomputer (with a seven-digit precision), we obtained 1.644897 (i.e., 1.6449 to four decimals). The use of 100 terms of the second series yielded 1.644888 (i.e., again 1.6449 to four decimals). We recall that the summation of 20,000 terms of the first series (double precision) gave us 1.644884, as obtained in the last section.

There seems to be no point in applying Kummer's method a third time because the gain achieved by further reduction in the number of terms does not compensate for the additional computational complexity of the terms of the new series, not to mention the laborious preparatory work. The laboratory instructor should point out, however, that even very laborious preparatory work is justified when the resulting formula is used a vast number of times, such as the formulas in computer library routines.

In summary, Kummer's method can be applied only if we can find a series with a known sum, whose behavior is similar to the series we want to sum. In other words, if we are interested in the sum $\sum a_n$, and we know that $\sum b_n = B$ and that $(a_n/b_n) \to 1$, then Kummer's method gives us

$$\sum_{n=1}^{\infty} a_n = B + \sum_{n=1}^{\infty} \left(1 - \frac{b_n}{a_n}\right) a_n = B + \sum_{n=1}^{\infty} \tilde{a}_n \tag{6.34}$$

The accelerated convergence of the series on the right hand-side of (6.34) stems from the fact that $(1 - b_n/a_n) \to 0$ and the faster this decays to zero, the more effective is Kummer's acceleration. In other words, the multiplication of a_n by the decaying factor $(1 - b_n/a_n)$ is the direct cause for the terms $\tilde{a}_n$ to decay to zero faster than the terms a_n. In this connection see Exercises 6.9 and 6.10. As a matter of fact, if $(a_n/b_n) \to C \neq 0$, where C is not necessarily equal to unity, we can still apply Kummer's method (see Exercise 6.11).

For series to which Kummer's method is not conveniently applicable, we can use other acceleration methods. One of those, which is based upon generalizations of geometric series, is presented next.

6.6 Approximate recursion relations

The acceleration method to be shown here is based on recursion relations, which are linear relations between any two or more consecutive terms of the series to be summed. Let us start by looking at geometric series and their generalizations.

The terms of an infinite geometric series $\sum a_n$ satisfy the relation $a_n = \alpha a_{n-1}$. If $|\alpha| < 1$, the sum is given by the closed form

$$S = \frac{a_0}{1-\alpha} = \frac{a_0}{1 - \dfrac{a_1}{a_0}} = \frac{a_0^2}{a_0 - a_1} \tag{6.35}$$

A closed-form formula can also be obtained for convergent series whose terms satisfy the three-term recursion relation

$$a_{n+1} = \alpha_1 a_n + a_2 a_{n-1} \tag{6.36}$$

This can be seen by summing both sides of (6.36) for $n = 1, 2, \ldots$ to get

$$S - a_0 - a_1 = \alpha_1 (S - a_0) + \alpha_2 S \tag{6.37}$$

and thus,

$$S = \frac{a_0 + a_1 - \alpha_1 a_0}{1 - \alpha_1 - \alpha_2} \tag{6.38}$$

If the values of α_1 and α_2 are known, the sum can be computed from (6.38). If not, α_1 and α_2 can be found from (6.36) for $n = 1, 2$. This yields

$$S = \frac{(a_0 + a_1)(a_1^2 - a_0 a_2) - a_0(a_2 a_1 - a_0 a_3)}{a_1^2 + a_2^2 + a_0(a_3 - a_2) - a_1(a_3 + a_2)} \tag{6.39}$$

We have thus expressed S in terms of a_0, a_1, a_2, and a_3. In general, if we have a series whose terms satisfy the recursion relation

$$a_{n+k-1} = \alpha_1 a_{n+k-2} + \cdots + \alpha_k a_{n-1} \tag{6.40}$$

for all n, then its sum can be expressed in terms of $a_0, a_1, \ldots, a_{2k-1}$. The resulting formulas, such as (6.39), are thus a natural extension of the geometric series formula (6.35). In this connection see Exercises 6.12 and 6.13.

We now make use of our ability to sum series whose terms satisfy a recursion relation of the type (6.40) to accelerate the convergence of a given series, whose terms satisfy such a relation approximately. The following example shows what is meant by "satisfy approximately." Let us consider the alternating series

$$\sum_{n=0}^{\infty} a_n = \sum_{n=0}^{\infty} \frac{(-1)^n}{2n+1} \tag{6.41}$$

To try to find a recursion relation between any three consecutive terms of (6.41), let us examine the following linear combination:

$$\begin{aligned} a_{n+1} - \alpha a_n - \beta a_{n-1} \\ &= \frac{(-1)^{n+1}}{2n+3} - \alpha\frac{(-1)^n}{2n+1} - \beta\frac{(-1)^{n-1}}{2n-1} \\ &= (-1)^n \frac{4(\beta - \alpha - 1)n^2 + 4(2\beta - \alpha)n + (3\beta + 3\alpha + 1)}{(4n^2 - 1)(2n+3)} \end{aligned} \tag{6.42}$$

We want to choose α and β so that the right-hand side vanishes, or at least becomes as small as possible, for large values of n. The requirement

$$\begin{aligned} \beta - \alpha &= 1 \\ 2\beta - \alpha &= 0 \end{aligned} \tag{6.43}$$

causes the coefficients of n^2 and n in the numerator to vanish and yields $\alpha = -2,\ \beta = -1$. Thus, we are left with

$$a_{n+1} + 2a_n + a_{n-1} = \frac{8(-1)^{n+1}}{(4n^2-1)(2n+3)} \tag{6.44}$$

The relation (6.44) is not an *exact* recursion relation because its right-hand side does not vanish. However, it is an *approximate* recursion relation in the sense that $|a_{n+1} + 2a_n + a_{n-1}|$ can be bounded by $1/n^3$, which is rather small for large n. This follows from

$$\begin{aligned} |a_{n+1} + 2a_n + a_{n-1}| &= \frac{8}{8n^3+12n^2-2n-3} \\ &< \frac{8}{8n^3+12n^2-2n^2-3n^2} \\ &= \frac{8}{8n^3+7n^2} < \frac{1}{n^3} \end{aligned} \tag{6.45}$$

where both inequalities result from the fact that decreasing the denominator increases the fraction. More generally, when we want to show that a recursion relation of the type (6.40) is satisfied approximately, we look for coefficients $\alpha_1, \alpha_2, \dots, \alpha_k$ such that

$$|a_{n+k-1} - \alpha_1 a_{n+k-2} - \cdots - \alpha_k a_{n-1}| < \frac{c}{n^p} \tag{6.46}$$

where p is as large as possible and c is an appropriate constant. In other words, we want to find the largest p such that the left-hand side of (6.46) is $O(1/n^p)$ [O was introduced following (6.13)]. Accordingly, the approximate recursion relation (6.44) states that $a_{n+1} + 2a_n + a_{n-1} = O(1/n^3)$. Other examples of approximate recursions are the subject of Exercises 6.14 and 6.15.

Returning to our series (6.41), whose sum S we want to find, let us sum the approximate recursion relation (6.44) for $n = 1, 2, \dots$ to obtain

$$(S - a_0 - a_1) + 2(S - a_0) + S = 8\sum_{n=1}^{\infty} \frac{(-1)^{n+1}}{(4n^2-1)(2n+3)} \tag{6.47}$$

Because $a_0 = 1$ and $a_1 = -\frac{1}{3}$, we find that

$$S = \sum_{n=0}^{\infty} \frac{(-1)^n}{2n+1} = \frac{2}{3} + 2\sum_{n=1}^{\infty} \frac{(-1)^{n+1}}{(4n^2-1)(2n+3)} \tag{6.48}$$

The method thus enables us to replace the original series by a series that converges faster, as will be shown. Both series in (6.48) alternate in sign, and thus by (6.10) their remainders satisfy

$$\left| \sum_{n=k+1}^{\infty} \frac{(-1)^n}{2n+1} \right| \le \frac{1}{2k+3} < \frac{1}{2k} \tag{6.49}$$

$$\left| 2 \sum_{n=k+1}^{\infty} \frac{(-1)^{n+1}}{(4n^2-1)(2n+3)} \right| \le \frac{2}{(4k^2+8k+3)(2k+5)} < \frac{1}{4k^3} \tag{6.50}$$

Now, to compute S so that a tolerance of $\frac{1}{2} \cdot 10^{-4}$ is assured, 10,000 terms are needed in the original series. On the other hand, 18 terms will suffice to assure this accuracy when we use the new series, on the right of (6.48), obtained through the acceleration method generated by the approximate recursion relation (6.44). In this connection see Exercise 6.16.

Our acceleration method can be carried further by examining four consecutive terms, rather than three, using the coefficients α, β, and γ. For our series (6.41) we thus obtain

$$\begin{aligned} &a_{n+2} - \alpha a_{n+1} - \beta a_n - \gamma a_{n-1} \\ &\quad = \frac{(-1)^n}{(2n+3)(2n+5)(4n^2-1)} \Big[8(\alpha - \beta + \gamma + 1)n^3 \\ &\quad + 4(5\alpha - 7\beta + 9\gamma + 3)n^2 + 2(-\alpha - 7\beta + 23\gamma - 1)n \\ &\quad + (-5\alpha + 15\beta + 15\gamma - 3) \Big] \end{aligned} \tag{6.51}$$

Using the same reasoning as before, we choose $\alpha = -3, \beta = -3$, and $\gamma = -1$, so that the coefficients of n^3, n^2, and n vanish. This leads to

$$a_{n+2} + 3a_{n+1} + 3a_n + a_{n-1} = \frac{48(-1)^{n+1}}{(2n+3)(2n+5)(4n^2-1)} = O\left(\frac{1}{n^4}\right) \tag{6.52}$$

Summing both sides for $n = 1, 2, \ldots$ and using also the values $a_0 = 1$, $a_1 = -\frac{1}{3}$, and $a_2 = \frac{1}{5}$, we end up with

$$S = \frac{11}{15} + 6 \sum_{n=1}^{\infty} \frac{(-1)^{n+1}}{(2n+3)(2n+5)(4n^2-1)} \tag{6.53}$$

Checking the remainder this time, we find that it is bounded by $0.375/k^4$, and thus only 10 terms are needed to assure the tolerance of $\frac{1}{2} \cdot 10^{-4}$.

The laboratory participants will observe that the additional preparatory work was not worthwhile, particularly because the terms of the new series are computationally more complex. The advantage of (6.53) over (6.48) would manifest itself only if much greater accuracy were required. Nevertheless, adding 10 terms of (6.53), we found $S = 0.7854$ to four decimals. Adding 18 terms in (6.48) yielded the same result. Just as a check, we summed 10,000 terms in the original series, using double precision, and also obtained 0.7854 to four decimals. We had to use double precision, for such large values of n, because the accuracy would otherwise have been damaged as round-off errors accumulated. The need to avoid such accumulations emphasizes the importance of acceleration methods. At this point see Exercises 6.17 and 6.18.

The approximate recursion method presented here is not limited to series whose terms alternate in sign, such as (6.41). To see this, consider the series

$$S = \sum_{n=0}^{\infty} a_n = \sum_{n=0}^{\infty} \frac{r^n}{(n+1)^2}, \qquad -1 \le r < 1 \tag{6.54}$$

To discover the approximate recursion relation for this series, we write

$$\begin{aligned} &a_{n+1} - \alpha a_n - \beta a_{n-1} \\ &\quad = r^{n-1}\big[(r^2 - \alpha r - \beta)n^4 + (2r^2 - 4\alpha r - 6\beta)n^3 \\ &\qquad + (r^2 - 4\alpha r - 13\beta)n^2 - 12\beta n - 4\beta\big]/\big[n^2(n+1)^2(n+2)^2\big] \end{aligned} \tag{6.55}$$

Choosing α and β so that the coefficients of n^4 and n^3 vanish, we have

$$\begin{aligned} \alpha r + \beta &= r^2 \\ 4\alpha r + 6\beta &= 2r^2 \end{aligned} \tag{6.56}$$

from which $\alpha = 2r$ and $\beta = -r^2$. Accordingly, we rewrite (6.55) in the form

$$a_{n+1} - 2ra_n + r^2 a_{n-1} = \frac{2r^{n+1}(3n^2 + 6n + 2)}{n^2(n+1)^2(n+2)^2} \tag{6.57}$$

A closer look at (6.57) reveals that its right-hand side is $O(1/n^4)$. Indeed, we have the approximate recursion relation

$$\begin{aligned} |a_{n+1} - 2ra_n + r^2 a_{n-1}| &< \frac{2|r|^{n+1}(3n^2 + 6n + 3)}{n^2(n+1)^2(n+2)^2} \\ &= \frac{6|r|^{n+1}}{n^2(n+2)^2} < \frac{6|r|^{n+1}}{n^4} \le \frac{6}{n^4} \end{aligned} \tag{6.58}$$

where the last inequality follows from $-1 \le r < 1$, stipulated in (6.54). Next we sum the relation (6.57) for $n = 1, 2, \ldots$ to obtain

$$(S - a_0 - a_1) - 2r(S - a_0) + r^2 S = 2\sum_{n=1}^{\infty} \frac{(3n^2 + 6n + 2)r^{n+1}}{n^2(n+1)^2(n+2)^2} \tag{6.59}$$

Because $a_0 = 1$ and $a_1 = r/4$, we find that

$$S = \sum_{n=0}^{\infty} \frac{r^n}{(n+1)^2} = \frac{1 - \frac{7}{4}r}{(1-r)^2} + \frac{2r}{(1-r)^2} \sum_{n=1}^{\infty} \frac{(3n^2 + 6n + 2)r^n}{n^2(n+1)^2(n+2)^2} \tag{6.60}$$

We have thus achieved the desired acceleration because the terms of the new series, on the right of (6.60), decrease faster than those of the original series, as is evident from (6.54) and (6.58). The most dramatic acceleration effect is obtained for $r = -1$, for which the remainder of the original series satisfies

$$\left| \sum_{n=k+1}^{\infty} \frac{(-1)^n}{(n+1)^2} \right| \le \frac{1}{(k+2)^2} < \frac{1}{k^2} \tag{6.61}$$

On the other hand, for the accelerated series we have

$$\begin{aligned} &\left| \frac{-2}{4} \sum_{n=k+1}^{n} \frac{(3n^2 + 6n + 2)(-1)^n}{n^2(n+1)^2(n+2)^2} \right| \\ &\qquad \le \frac{2\left[3(k+1)^2 + 6(k+1) + 2\right]}{4(k+1)^2(k+2)^2(k+3)^2} < \frac{6}{4(k+1)^4} < \frac{1.5}{k^4} \end{aligned} \tag{6.62}$$

where (6.57) and (6.58) have been invoked. Now, to compute S so that a tolerance of $\frac{1}{2} \cdot 10^{-6}$ is assured, 1415 terms of the original series are needed compared with a mere 42 terms of the new, accelerated series. For $r = -0.9$, for example, the acceleration effect is far less dramatic, and it turns out that 23 terms, rather than 60 in the original series, are needed for this accuracy.

In the preceding example, which is actually a power series in disguise, we excluded the case $r = 1$, for which the method breaks down. This can be seen by substituting $r = 1$ in (6.59) and observing that S drops out. Fortunately, this case is tailor-made for Kummer's acceleration method, as can be seen from (6.28) in which n is replaced by $(n + 1)$.

The convergence of our power series (6.54) can be further accelerated by using four consecutive terms rather than three, as was done for $\sum(-1)^n/(2n + 1)$. In this connection see Exercise 6.19. Another example of a power series whose convergence can be accelerated by using approximate recursion is $S = \sum r^n/(n+1)$, $-1 \leq r < 1$, as given in Exercises 6.20, 6.21, and 6.22. We note that the convergence acceleration constructed in this section is valid not only for $-1 \leq r < 0$ but also for $0 < r < 1$, that is, when the terms of the series are positive and do not alternate in sign. This can be seen from (6.60), for example, in which the coefficient of r^n in the original series decreases like $(n+1)^2$ whereas the corresponding coefficient in the accelerated series decreases essentially like $(n+1)^4$. So far, the remainder R_k has been estimated only for $-1 \leq r < 0$. We presently show how to estimate it for $0 < r < 1$.

Reconsider the infinite series $\sum r^n/(n+1)^2$ for positive r and its remainder

$$R_k = \sum_{n=k+1}^{\infty} \frac{r^n}{(n+1)^2}, \qquad 0 < r < 1 \tag{6.63}$$

which can be estimated by

$$0 < R_k \leq \sum_{n=k+1}^{\infty} \frac{r^n}{n(n+1)} \tag{6.64}$$

We recall that the decomposition $1/[n(n+1)] = 1/n - 1/(n+1)$ enabled us earlier to sum series that can thus be written in telescoping form. The series in (6.64) is not immediately amenable to telescoping, and so we employ the following artifice, which will produce the desired form. Since $0 < r < 1$, we have

$$R_k \leq \sum_{n=k+1}^{\infty} \frac{r^n}{n(n+1)} \leq \sum_{n=k+1}^{\infty} \frac{[1 + n(1-r)]r^n}{n(n+1)} \tag{6.65}$$

because the numerator is increased by the factor $F = 1 + n(1-r) = (n+1) - nr > 1$. Thus, we can rewrite (6.65) in the form

$$R_k \leq \sum_{n=k+1}^{\infty} \frac{(n+1) - nr}{n(n+1)} r^n = \sum_{n=k+1}^{\infty} \left[\frac{r^n}{n} - \frac{r^{n+1}}{n+1}\right] \tag{6.66}$$

which telescopes to $r^{k+1}/(k+1)$. Thus,

$$0 < R_k \leq \frac{r^{k+1}}{k+1} < \frac{r^{k+1}}{k} \tag{6.67}$$

To estimate the remainder of the corresponding accelerated series in (6.60), we proceed analogously. Thus,

$$\begin{aligned} 0 < \tilde{R}_k &= \frac{2r}{(1-r)^2} \sum_{n=k+1}^{\infty} \frac{(3n^2+6n+2)r^n}{n^2(n+1)^2(n+2)^2} \\ &\le \frac{2r}{(1-r)^2} \sum_{n=k+1}^{\infty} \frac{3r^n}{n^2(n+2)^2} \\ &\le \frac{6r}{(1-r)^2} \sum_{n=k+1}^{\infty} \frac{r^n}{(n-1)n(n+1)(n+2)} \end{aligned} \tag{6.68}$$

This time the factor F, by which we must multiply the numerator to obtain telescoping, turns out to be

$$F = \frac{1}{3}\left[(n+2)-(n-1)r\right] = 1 + \frac{n-1}{3}(1-r) > 1 \tag{6.69}$$

and we have

$$\begin{aligned} \tilde{R}_k &\le \frac{2r}{(1-r)^2} \sum_{n=k+1}^{\infty} \frac{(n+2)-(n-1)r}{(n-1)n(n+1)(n+2)} r^n \\ &= \frac{2r}{(1-r)^2} \sum_{n=k+1}^{\infty} \left[\frac{r^n}{(n-1)n(n+1)} - \frac{r^{n+1}}{n(n+1)(n+2)}\right] \\ &= \frac{2r}{(1-r)^2} \cdot \frac{r^{k+1}}{k(k+1)(k+2)} \le \frac{2r^{k+2}}{(1-r)^2 k^3} \end{aligned} \tag{6.70}$$

Note that the factor F in (6.69) is of the form

$$F = \frac{(n+i)-(n-j)r}{i+j} = 1 + \frac{n-j}{i+j}(1-r) > 1 \tag{6.71}$$

where $(n+i)$ is the largest factor, and $(n-j)$ the smallest, of the denominators in (6.68). The same is true for the factor F incorporated in (6.65) and (6.66). For the generalization of this idea see Exercise 6.23.

Now we want to assess the efficiency of the acceleration – say, for $r = 0.9$ – when we are interested in the sum of the series correct to six decimal figures. Summing the original series $\sum r^n/(n+1)^2$ and using the estimate (6.67), we find that 94 terms are needed to assure this accuracy. On the other hand, summing the accelerated series in (6.60) and using the estimate (6.70), we see that 67 terms suffice. The gain achieved here is even more modest than for $r = -0.9$, where the number of terms was reduced from 60 to 23. This, however, is the case because of the denominator $(1-r)^2$ in the accelerated series, which has a magnifying effect for $r = +0.9$. The result for $r = 0.9$, correct to six decimal figures, turned out to be $S = 1.444127$.

In this laboratory assignment we have discussed methods of accelerating the convergence of infinite series by means of Kummer's method, based on related series whose sum can be expressed in closed form, and a method based on approximate recursion relations among a few consecutive terms of the series. Typical examples have been given along with actual computations that show the increase in the rate of convergence and the corresponding efficiency of the computation. We have not only shown how to improve computational efficiency but also have emphasized the difference between a theoretical proof that a series converges and the actual computation of its sum to a given accuracy.

Exercises

6.1 Show that (a) $\alpha k^2 + \beta k + \gamma = O(k^2)$, (b) $A\sqrt{k} + B\sqrt[3]{k} = O(\sqrt{k})$, (c) $D/(2k+9)^2 = O(1/k^2)$, (d) $k^2/(7+\sqrt{k}) = O(k^{3/2})$.

6.2 Show that (a) $\sin k = o(k)$, (b) $1/k! = o(2^{-k})$, (c) $2^{-k} = o(1/k^2)$. [Hint: examine the factor by which the appropriate fraction is multiplied when passing from k to $(k+1)$.]

6.3 Recall that the sum S of an infinite geometric series $a_1 + a_1 r + a_1 r^2 + \cdots$, with $|r| < 1$, is given by $S = a_1/(1-r)$. Next, find a simple fraction that equals (a) $0.2222\ldots$, (b) $3.17171717\ldots$, (c) $0.9358585858\ldots$.

6.4 Choose $S = 2$ and construct infinite series converging to S as in (6.16). Use (a) $x_n = 1/(n+1)^2$, (b) $x_n = 1/\sqrt{n+1}$.

6.5 Occasionally, a given series can be rewritten as a telescoping series, which makes it possible to find its sum. Apply this method to

$$S = \sum_{n=1}^{\infty} \frac{1}{n(n+1)(n+2)}$$

6.6 Find upper and lower bounds for the remainder after k terms of the series $\sum(1/n^3)$. Use these bounds to construct a correction term, to be added to the first k terms, such that a smaller value of k guarantees a given accuracy. How many terms are needed before and after the addition of the correction term, respectively, so that six correct figures are guaranteed.

6.7 Repeat Exercise 6.6 for $\sum(1/n^4)$ so that twelve correct figures are guaranteed.

6.8 Write and run computer programs for the summation of $\sum(1/n^2)$, and print out results every 200 terms. Sum 20,000 terms altogether, and note at what point the first four figures stop changing. Repeat the computation using double precision. Do the results agree with those reported in the text. If not, explain.

6.9 Apply Kummer's acceleration method to the series

$$S = \sum_{n=3}^{\infty} \frac{n^2}{n^4 + 1}$$

How many terms are needed before and after the acceleration to guarantee $q = 4$ correct decimal figures. Apply the method once again, and assess the additional improvement.

6.10 Repeat Exercise 6.9 for the series

$$S = \sum_{n=1}^{\infty} \frac{1}{n^3}$$

with $q = 6$. Write and run corresponding computer programs to obtain the results for all three series.

6.11 Modify Kummer's method to accelerate the convergence of $\sum a_n$ with the aid of $\sum b_n = B$, given that $(a_n/b_n) \to C \neq 0$.

6.12 Show that any three consecutive terms of the series

$$S = \sum_{n=0}^{\infty} a_n = \sum_{n=0}^{\infty} \frac{n+1}{2^n}$$

satisfy a recursion relation of the type $a_{n+1} = \alpha a_n + \beta a_{n-1}$. Now find S by summing both sides of this relation for $n = 1, 2, 3, \ldots$.

6.13 Repeat Exercise 6.12 with $a_n = (n^2+3)/5^n$. This time, however, the underlying property is to be a four-term recursion relation.

6.14 Find coefficients α and β such that the terms of the series

$$S = \sum_{n=0}^{\infty} a_n = \sum_{n=0}^{\infty} \frac{(-1)^n}{n+1}$$

satisfy an approximate three-term recursive relation, and express it in terms of the order-of-magnitude symbol O. Also, find c so as to rewrite the relation in the form (6.46).

6.15 Repeat Exercise 6.14 using α, β, and γ to generate an approximate four-term recursion relation. In addition, generate an approximate two-term recursion relation, using only one parameter.

6.16 Use the approximate three-term recursion relation obtained in Exercise 6.14 to accelerate the convergence of $\sum(-1)^n/(n+1)$. How many terms are needed to assure q decimal figures in the original and the accelerated series, respectively, for (a) $q = 4$ and (b) $q = 8$.

6.17 Use the approximate four-term recursion relation obtained in Exercise 6.15 to accelerate the convergence of $\sum(-1)^n/(n+1)$. How many terms are needed this time to assure $q = 4$ and $q = 8$ decimal figures, respectively. Compare the results with those of Exercise 6.16.

6.18 Write and run computer programs to find the sum of the series $S = \sum(-1)^n/(n+1)$ correct to four decimal figures, using this series as is and then its accelerated versions from Exercises 6.16 and 6.17.

6.19 Derive an approximate four-term recursion relation for the power series

$$\sum_{n=0}^{\infty} a_n = \sum_{n=0}^{\infty} \frac{r^n}{(n+1)^2}, \qquad -1 \leq r < 1$$

and proceed to find the corresponding accelerated series. How many terms are needed in this accelerated series, for $r = -1$ and $r = -0.9$, to assure six decimal figures. Compare with the results achieved in the text, by means of (6.60), through an approximate three-term recursion relation, for $r = -1$ and $r = -0.9$.

6.20 Derive an approximate three-term recursion relation in order to accelerate the convergence of the power series

$$\sum_{n=0}^{\infty} a_n = \sum_{n=0}^{\infty} \frac{r^n}{n+1}, \qquad -1 \leq r < 1$$

Determine the number of terms required to assure four decimal figures for $r = -1$ and $r = -0.9$, both before and after the acceleration.

6.21 Repeat Exercise 6.20 with an approximate four-term recursion relation.

6.22 Write and run computer programs for the sum $S = \sum r^n/(n+1)$ correct to four decimal figures, using this series as is and then its accelerated versions from Exercises 6.20 and 6.21. Use the value $r = -0.9$ as in Exercises 6.20 and 6.21, as well as a new value $r = -0.99$.

6.23 In the power series

$$S = \sum_{n=0}^{\infty} \frac{r^n}{(n+1)^p}$$

$p > 1$ is a positive integer. Estimate the remainder R_k for $0 < r < 1$ by introducing an appropriate factor F in the numerator and reaching a telescoping series.

7
Interpolative approximation

7.1 Introduction

In the spirit of the mathematical laboratory, special attention is accorded to the elimination of traditional mathematical tables, which students have always used as "black boxes," without the faintest understanding of their origin and construction. Thus, these tables were antieducational tools that the students aquired without any mathematical enlightenment and used as "cookbook recipes." Now that calculators and microcomputers are used in mathematical education, the danger arises of replacing one set of black boxes with another. Of course, we are not advocating the introduction of these new electronic black boxes merely for using, say, the logarithmic built-in function of the computer (or pressing the "log" key on a pocket calculator). What we do advocate is to teach students what is behind such built-in functions as part of the material covered in the mathematical laboratory. This subject fits naturally into the environment of the laboratory and reveals the "story behind the key." The attainment of this objective is the subject of this chapter and the next.

We might ask whether students should be allowed to use the built-in functions before (and during) learning how they were built in. We feel that no harm can result from such a practice, so long as the students are told expressly that their "ignorant" use of the built-in functions is *temporary*. Before long, the secrets held by the computer keys will be revealed. It is our experience that this revelation fully satisfies the curiosity of the students to know the answer to the tantalizing question: When we instruct a computer to evaluate $\log[\sin(\pi/10)]$, say, how does it supply the answer so accurately with such tremendous speed?!

7.2 Approximation by interpolation

Let us suppose we are faced with the problem of evaluating a function $f(x)$ for any x in the interval $[a, b]$. We assume that we know the values of $y = f(x)$ at $(n+1)$ specific points $x_0, x_1, x_2, \ldots, x_n$ in that interval. The points (x_j, y_j), where $j = 0, 1, 2, \ldots, n$, will later be used to construct a polynomial that serves as an efficient approximation of $f(x)$ in $[a, b]$. It is best, in the laboratory, to proceed with a concrete example that is familiar to the students, say $f(x) = \sin x$. The interval $[a, b]$ is $[0, \pi/2]$, and the ability to evaluate $y = \sin x$ everywhere in that interval furnishes us with the values of $\sin x$ for all x (in radians) by means of the usual trigonometric identities.

For definiteness we choose $n = 6$ (i.e., seven points), $x_0 = 0$ so that $y_0 = \sin x_0 = 0$, and $x_6 = \pi/2$ so that $y_6 = \sin \pi/2 = 1$. The remaining five points, x_1 through x_5, are distributed evenly over the interval $[0, \pi/2]$, in the absence (at this stage) of any motivation to do otherwise. Accordingly, we now have $x_1 = \pi/12$, $x_2 = \pi/6$, $x_3 = \pi/4$, $x_4 = \pi/3$, $x_5 = 5\pi/12$, and the corresponding values of y_1 through y_5 can be obtained. Thus, $y_2 = \sin \pi/6 = \sin 30°$ and $y_4 = \sin \pi/3 = \sin 60°$ are available from the $30°\text{–}60°\text{–}90°$ triangle, and $y_3 = \sin \pi/4 = \sin 45°$ is known from the isosceles right triangle. Furthermore, $y_1 = \sin \pi/12 = \sin 15°$ can be obtained by using the half-angle formulas. Indeed, we see that $y_1 = \sqrt{(1 - \cos 30°)/2} = \sqrt{2 - \sqrt{3}}/2$. Finally, $y_5 = \sin 5\pi/12 = \sin 75° = \cos 15° = \sqrt{(1 + \cos 30°)/2} = \sqrt{2 + \sqrt{3}}/2$. In this way we obtain the following seven pairs of values (correct to 10 decimal figures):

$$
\begin{array}{ll}
x_0 = 0.0000000000 & y_0 = 0.0000000000 \\
x_1 = 0.2617993878 & y_1 = 0.2588190451 \\
x_2 = 0.5235987757 & y_2 = 0.5000000000 \\
x_3 = 0.7853981635 & y_3 = 0.7071067813 \\
x_4 = 1.0471975513 & y_4 = 0.8660254039 \\
x_5 = 1.3089969392 & y_5 = 0.9659258263 \\
x_6 = 1.5707963270 & y_6 = 1.0000000000
\end{array}
\tag{7.1}
$$

One should not lose sight of the fact that the computations of π and various square roots, inherent in the displayed values, constitute by themselves typical subjects for the mathematical laboratory and have been discussed at length in Chapters 5 and 2, respectively. In addition, the use of standard trigonometric identities enables us to add more points to the seven already given, if we so desire (see Exercise 7.1). However, we confine the present discussion to the seven specific points displayed (i.e., $n = 6$), which will henceforward be called *nodes*.

Recall that our objective is to evaluate the function $y = \sin x$ when its values at the points $x_0, x_1, \ldots, x_6$ are known. The simplest procedure toward attaining this goal is to approximate $\sin x$ by a polygonal line, composed of line segments joining successive pairs of nodes. The equation of the segment joining (x_j, y_j) and (x_{j+1}, y_{j+1}) is found to be

$$P_1(x) = \frac{x - x_{j+1}}{x_j - x_{j+1}} y_j + \frac{x - x_j}{x_{j+1} - x_j} y_{j+1}, \qquad x_j \leq x \leq x_{j+1} \qquad (7.2)$$

Equation (7.2) holds of course for $j = 0, 1, \ldots, 5$, and the notation $P_1(x)$ refers to the fact that (7.2) represents, for each j, a segment of a straight line, that is, a polynomial of the first degree. With a view toward later generalizations of (7.2), let us observe that for $x = x_j$ the second term on the right of (7.2) vanishes whereas the coefficient of y_j in the first term equals unity, and thus $P_1(x_j) = y_j$. Analogously, $P_1(x_{j+1}) = y_{j+1}$. Clearly, six expressions of the type (7.2), one for each subinterval, represent the required polygonal approximation to $\sin x$.

An equation of the type (7.2) is called an *interpolation formula* because it furnishes approximate intermediate values of the function we want to evaluate. Thus, the approximation generated by (7.2) constitutes a piecewise linear interpolation.

While we carry out the procedure in the mathematical laboratory, we should consider the possibility of using the data in (7.1) to construct a better approximation to $\sin x$ by a method that yields a tighter local fit. The most natural thing to do is to take the nodes three at a time and pass a parabola through each triplet of points. In our case we would have three such triplets, each of the form (x_{j-1}, y_{j-1}), (x_j, y_j), (x_{j+1}, y_{j+1}), for $j = 1, 3, 5$. The equation of such a parabola should be an extension of (7.2), and a unique parabola passes through a given triplet of points (two distinct such parabolas would imply three distinct roots of a quadratic equation; see Exercise 7.2). To obtain the desired extension of formula (7.2), we use a sum of three terms corresponding to the three values of y, such that each coefficient is a quotient of quadratic rather than linear expressions. Moreover, each coefficient should vanish for two of the points and equal unity for the third. Thus, we are led to

$$P_2(x) = \frac{(x - x_j)(x - x_{j+1})}{(x_{j-1} - x_j)(x_{j-1} - x_{j+1})} y_{j-1} + \frac{(x - x_{j-1})(x - x_{j+1})}{(x_j - x_{j-1})(x_j - x_{j+1})} y_j$$
$$+ \frac{(x - x_{j-1})(x - x_j)}{(x_{j+1} - x_{j-1})(x_{j+1} - x_j)} y_{j+1}, \quad x_{j-1} \leq x \leq x_{j+1} \qquad (7.3)$$

Each term is of second degree and hence so is the sum, which is therefore denoted by $P_2(x)$. As a check, if we substitue $x = x_{j-1}$, say, we find that the coefficient of y_{j-1} equals unity and the two others vanish; hence $P_2(x_{j-1}) = y_{j-1}$. Similarly, we find $P_2(x_j) = y_j$ and $P_2(x_{j+1}) = y_{j+1}$. Summing, we can now construct three consecutive parabolic arcs, each of the form (7.3), which together give us a piecewise quadratic interpolation for $\sin x$.

Having gone this far, we wonder how to extend this approximaton method even beyond quadratic interpolation. Could we pass *two* successive cubic polynomials through the first four nodes and the last four nodes, respectively? This would lead to the following general question: why not pass one polynomial of nth degree through all $(n+1)$ nodes and use it to approximate the function $f(x)$? (In our case $f(x) = \sin x$ and $n = 6$.) It turns out that such a polynomial does exist and, moreover, it is unique (see Exercise 7.3). This polynomial, denoted $P_n(x)$, is referred to as the interpolating polynomial of degree n (coinciding with the given $f(x)$ at the nodes), where n is one less than the number of given nodes. Thus, in our case $y = \sin x$ with seven nodes, we eventually seek the polynomial of degree six that will do the job.

In the general case, it seems reasonable in view of (7.3) and (7.2) to look for an nth-degree polynomial of the form

$$P_n(x) = \sum_{j=0}^{n} \frac{N_j(x)}{D_j} y_j \tag{7.4}$$

where $N_j(x)$ is an nth-degree polynomial such that $N_j(x_j)/D_j = 1$ and $N_j(x_k) = 0$ for $k \neq j$. If we look again at (7.3) and (7.2), we can conclude that

$$\frac{N_j(x)}{D_j} = \frac{(x - x_0) \cdots (x - x_{j-1})(x - x_{j+1}) \cdots (x - x_n)}{(x_j - x_0) \cdots (x_j - x_{j-1})(x_j - x_{j+1}) \cdots (x_j - x_n)} \tag{7.5}$$

We verify that (7.5) indeed represents an nth-degree polynomial with the required properties, because the factor $(x - x_j)$ is missing in the numerator. For $j = 0$ and $j = n$, the first and last factors, respectively, of numerator (and denominator) are the missing ones. The interpolation polynomial given by (7.4) and (7.5) is due to J. L. Lagrange (1736–1813) and bears his name.

For computational purposes – when $(n + 1)$ specific nodes are given – we can rewrite (7.4)–(7.5), after preliminary algebraic manipulations, in the form

$$P_n(x) = a_n x^n + a_{n-1} x^{n-1} + \cdots + a_1 x + a_0 \tag{7.6}$$

Here the coefficients $a_0, a_1, \dots, a_n$ are expressed in terms of the given x_j and y_j (see Exercise 7.4). Although this preliminary work is considerable, the student should realize that it is done only once. Thus, the use of (7.6) in repeated computations, to approximate $f(x)$ for various values of x, is computationally far more efficient than the use of (7.4) and (7.5). We come back to this point in Chapter 8 in connection with computer library functions.

So far, we have not discussed the cardinal question of the quality of the approximation we can expect from $P_n(x)$. That is to say, when $P_n(x)$ is used as an approximation to $f(x)$ for various values of x in $[a, b]$, what errors are incurred and what can be said about their magnitude?

7.3 Quality of approximation

We are now ready to use the seven nodes in (7.1) to construct in the mathematical laboratory the polynomial $P_6(x)$ that approximates $\sin x$ in $[0, \pi/2]$. We shall then evaluate $P_6(x)$ for values of x increasing from $x = 0$ to $x = \pi/2$, with increments of $\pi/180$, say, corresponding to increments of one degree. The values thus obtained are then compared with the corresponding values of $\sin x$ given by the computer's built-in sine function, and the differences

$$R_6(x) = \sin x - P_6(x) \tag{7.7}$$

are recorded and plotted. The function $R_6(x)$ is referred to as the remainder, and it should be plotted on an appropriate scale that accentuates its behavior and thereby the quality of approximations. We actually carried out the computations indicated and found experimentally that

$$|R_6(x)| = |\sin x - P_6(x)| < 1.3 \cdot 10^{-6}, \qquad 0 \le x \le \frac{\pi}{2} \tag{7.8}$$

By *experimentally* we mean that the error bound $1.3 \cdot 10^{-6}$ has not been deduced analytically but is a result of the laboratory computations just described. These experimental results indicate that a sixth-degree polynomial, based on just seven given nodes, furnishes us with accuracy between five and six decimal figures. We verified the claim made in (7.8) by repeating these computations in the mathematical laboratory, using increments of $\pi/360$ as well as $\pi/720$. The use of yet smaller increments is recommended (see Exercise 7.5) so that the laboratory participants get an even better feeling of the quality of approximation achieved by $P_6(x)$.

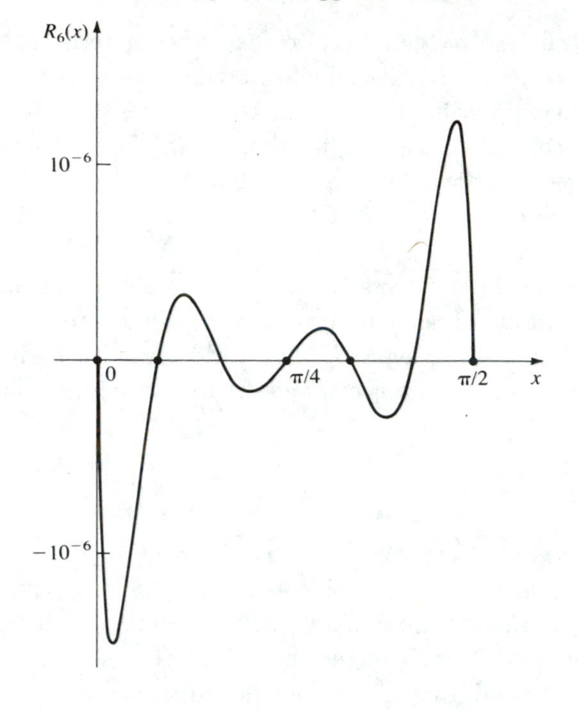

Fig. 7.1

To see the behavior of the remainder $R_6(x)$ in a vivid way, we plotted it on a set of axes with a stretched y-scale and obtained Fig. 7.1. The accuracy obtained by approximating $\sin x$ with $P_6(x)$ is so good that, had we plotted $\sin x$ and $P_6(x)$ simultaneously, the two graphs would have been virtually identical and, in any case, practically indistinguishable. It is only when we plot the stretched remainder $R_6(x)$ that we can see it clearly and study its behavior. To truly appreciate the high quality of approximation generated by our $P_6(x)$, we recommend the construction of $P_3(x)$, placing the nodes at $x = 0,\ \pi/6,\ \pi/3,\ \pi/2$, and plotting the corresponding $R_3(x)$. In this connection see Exercises 7.6 and 7.7. The nodes, also known as *interpolation points*, need not be equally spaced nor need they include the interval's endpoints.

We want to underscore the fact that $P_6(x)$ is compared with the computer's built-in sine function, which itself is based on some approximation. However, because the built-in sine function has an accuracy of, say, twelve decimal figures, its comparison with $P_6(x)$ can be regarded as the comparison of $P_6(x)$ with the true sine function. Moreover, this comparison sheds light on the methods by which built-in functions can be constructed, as will be discussed in Chapter 8. However, if the computer's built-in functions are not highly accurate, we must be content with $P_3(x)$ or $P_4(x)$ as an approximation to $\sin x$ and with the corresponding lower quality of approximation. If we use too many interpolation points (i.e., nodes), the resulting polynomial might be more accurate than the computer's built-in function, which can then no longer be regarded as the true sine function. Moreover, under these circumstances round-off errors can become dominant and render the results meaningless.

We turn now to a closer study of the behavior of $R_6(x)$ in Fig. 7.1, particularly the size of $\text{Max}|R_6(x)|$ in the underlying interval:

1. $R_6(x)$ vanishes, naturally, at our seven interpolation points. Elsewhere, $R_6(x)$ oscillates between positive and negative values, reflecting the fact that $P_6(x)$ winds and wraps itself around $\sin x$.
2. If we were to add two more nodes, placing them near the interval endpoints, such as $x = \pi/24$ and $x = 11\pi/24$ (corresponding to $7.5°$ and $82.5°$), the error $|R_6(x)|$ would tend to decrease where the graph shows its magnitude to be larger.
3. We could redistribute our seven interpolation points if we must do with just seven, shifting the second (and perhaps the third) node toward the left endpoint, and the sixth (and fifth) node toward the right. The purpose is to "smear" the error as uniformly as possible throughout the interval, thus decreasing the magnitude of its maxima.
4. If the situation described for $\sin x$ is any indication of what happens in general (advanced methods show that it is), then we should choose a higher density of nodes toward the interval's endpoints and a lower density around its center.
5. We stress that the attempt to smear the error uniformly over the interval does not require placing the first and last nodes at the endpoints of the interval. Allowing the positions of the first and last nodes to vary can reduce the remainder's maxima even more.

At this juncture the laboratory participants should be encouraged to experiment by trial and error with the location of the nodes, given their number. A few more nodes can also be added, but their number should be limited. The guiding principle is that efficiency be maintained. That is to say, the attainment of greater accuracy must not be offset by unreasonable, additional computational effort caused by too many nodes. For example, it is certainly unreasonable to use an approximating polynomial of degree 100 to obtain very accurate values of $\sin x$.

This section serves as a point of departure for a number of profound mathematical questions, whose full treatment is beyond our scope. We mention a few of them: If we have a function $f(x)$ over an interval $[a, b]$ that we want to approximate by an interpolation polynomial $P_n(x)$ for a given n, then we can ask the following:

1. Do there exist optimal nodes, in the sense that the maximum value of $|R_n(x)| = |f(x) - P_n(x)|$ for all x in the interval is minimized? In other words, can we minimize the maximum deviation of $P_n(x)$ from $f(x)$?
2. If such an optimal polynomial exists, it it unique?
3. If it exists and is unique, does this optimal polynomial possess characteristic properties leading to some procedure by which it can be constructed?

The Russian mathematician P. Chebyshev showed that under rather general conditions the answers to these questions are affirmative. This polynomial is called the corresponding *minimax polynomial* because it minimizes the maximum deviation, and, indeed, it oscillates about $f(x)$ in a way that smears the error uniformly (*equal ripple property*). These results can be verified in the mathematical laboratory for a very simple case: the approximation of a convex (concave) curve by a straight line. In this connection see Exercise 7.8 and Section 2.5 where the straight line $P_1(x) = x/3 + 17/24$ is actually the minimax line corresponding to $y = \sqrt{x}$ in $[1, 4]$.

In this section we have approximated $y = \sin x$ by an interpolative polynomial $P_6(x)$ and have studied the behavior of the remainder $R_6(x) = \sin x - P_6(x)$. This is but a special case because we could equally well have constructed an interpolative approximation $P_n(x)$ for other functions such as $y = \ln x$ in a suitable interval. Moreover, our approximation of sin x turned out to be of a very high quality, but its construction did not reveal why such high quality could be expected. In other words, estimation of the quality of approximation is still a mystery to the laboratory participants, as is the problem of constructing a reliable built-in function without using an existing built-in function (as we have done so far). Shedding light on these issues requires a closer look at the remainder $R_n(x) = f(x) - P_n(x)$, to which we turn our attention in the next section.

7.4 A closer look at $R_n(x)$

To study the quality of interpolative approximation, we restrict ourselves (to begin with) to first- and second-order polynomials of interpolation. We also pave the way for generalizations to higher order.

Suppose we want to approximate $y = f(x)$ in $[a, b]$, knowing the values $y_0 = f(x_0)$ and $y_1 = f(x_1)$, where x_0 and x_1 are two distinct points in $[a, b]$. If $P_1(x)$ denotes the first-degree polynomial passing through (x_0, y_0) and (x_1, y_1), then

$$P_1(x) = y_0 + \frac{y_1 - y_0}{x_1 - x_0}(x - x_0) \tag{7.9}$$

Obviously, this is but a different form of the Lagrange representation (7.2), which in our case here would be

$$P_1(x) = \frac{x - x_1}{x_0 - x_1} y_0 + \frac{x - x_0}{x_1 - x_0} y_1 \tag{7.10}$$

The use of (7.9) and its generalizations was suggested by Newton as a means of studying the remainder $R_n(x)$ because it is superior to Lagrange's form, (7.4) and (7.5), for that purpose. Indeed, if a third point (x_2, y_2) is given, with x_2 in $[a, b]$, we can construct the second-degree polynomial $P_2(x)$ that coincides with $f(x)$ at the three interpolation points, as follows:

$$P_2(x) = P_1(x) + \gamma \cdot (x - x_0)(x - x_1) \tag{7.11}$$

where γ is chosen so that $P_2(x_2) = y_2$. Because $P_2 = P_1$ at x_0 and x_1, the collocation of P_2 with our function at these points is guaranteed. In a similar way, we can construct $P_3(x)$ when a fourth point is given, and so on. The requirement $P_2(x_2) = y_2$ yields (see Exercise 7.9),

$$\gamma = \frac{y_2 - P_1(x_2)}{(x_2 - x_0)(x_2 - x_1)} = \frac{\frac{y_2 - y_0}{x_2 - x_0} - \frac{y_1 - y_0}{x_1 - x_0}}{x_2 - x_1} \tag{7.12}$$

Because the quantity on the right-hand side of (7.12) will be used several times throughout this computational assignment, we introduce the notation

$$\gamma(f; x_0, x_1, x_2) = \frac{\frac{y_2 - y_0}{x_2 - x_0} - \frac{y_1 - y_0}{x_1 - x_0}}{x_2 - x_1} \tag{7.13}$$

and thus

$$P_2(x) - P_1(x) = \gamma(f; x_0, x_1, x_2) \cdot (x - x_0)(x - x_1) \tag{7.14}$$

Now if we substitute $x = x_2$ and recall that $P_2(x_2) = f(x_2)$, we obtain

$$f(x_2) - P_1(x_2) = \gamma(f; x_0, x_1, x_2) \cdot (x_2 - x_0)(x_2 - x_1) \tag{7.15}$$

Because x_2 could be any point in the interval $[a, b]$, we can rewrite (7.15) in the form

$$f(x) - P_1(x) = \gamma(f; x_0, x_1, x) \cdot (x - x_0)(x - x_1) \tag{7.16}$$

Equality (7.16) forms the basis for estimating the quality of the polynomial approximation of f in $[a, b]$ because $[f(x) - P_1(x)]$ is actually the remainder $R_1(x)$, that is,

$$R_1(x) = f(x) - P_1(x) = \gamma \cdot (x - x_0)(x - x_1) \tag{7.17}$$

Prior to the study of such an estimate, it is expedient to derive the extension of (7.9) to a quadratic polynomial $P_2(x)$ that coincides with $f(x)$ at three points (x_0, y_0), (x_1, y_1), and (x_2, y_2). If we substitute (7.12) into (7.11), we find

$$P_2(x) = y_0 + \frac{y_1 - y_0}{x_1 - x_0}(x - x_0) + \frac{\frac{y_2 - y_0}{x_2 - x_0} - \frac{y_1 - y_0}{x_1 - x_0}}{x_2 - x_1}(x - x_0)(x - x_1) \tag{7.18}$$

We note that the interpolation parabola (7.18) is but a different (Newton's) form of the Lagrange representation (7.3), which in our case here would be

$$\begin{aligned} P_2(x) = {} & \frac{(x - x_1)(x - x_2)}{(x_0 - x_1)(x_0 - x_2)} y_0 + \frac{(x - x_0)(x - x_2)}{(x_1 - x_0)(x_1 - x_2)} y_1 \\ & + \frac{(x - x_0)(x - x_1)}{(x_2 - x_0)(x_2 - x_1)} y_2 \end{aligned} \tag{7.19}$$

In this connection see Exercise 7.10. The advantage of Newton's over Lagrange's form manifests itself in the following: When we have an interpolation polynomial $P_{n-1}(x)$ based on n interpolation points and we want to construct $P_n(x)$ such that it passes also through an additional interpolation point, then Lagrange's form (7.4) and (7.5) necessitates the construction of the entire polynomial $P_n(x)$ from scratch. On the other hand, Newton's form enables us to construct $P_n(x)$ by adding one appropriate term to the previously constructed $P_{n-1}(x)$. This property also makes it possible to determine the structure of the remainder, as we did with $R_1(x)$ and will do with $R_2(x)$.

To find the remainder $R_2(x) = f(x) - P_2(x)$, we construct $P_3(x)$, which should pass through (x_0, y_0), (x_1, y_1), (x_2, y_2), and a fourth distinct point (x_3, y_3). Thus, we search for δ, satisfying

$$P_3(x) = P_2(x) + \delta \cdot (x - x_0)(x - x_1)(x - x_2) \tag{7.20}$$

as well as

$$P_3(x_3) = y_3 \tag{7.21}$$

This leads to

$$\begin{aligned}\delta &= \frac{y_3 - P_2(x_3)}{(x_3 - x_0)(x_3 - x_1)(x_3 - x_2)} \\ &= \frac{\frac{1}{x_3-x_1}\left[\frac{y_3-y_0}{x_3-x_0} - \frac{y_1-y_0}{x_1-x_0}\right] - \frac{1}{x_2-x_0}\left[\frac{y_2-y_1}{x_2-x_1} - \frac{y_1-y_0}{x_1-x_0}\right]}{x_3 - x_2} \\ &= \delta(f; x_0, x_1, x_2, x_3)\end{aligned} \tag{7.22}$$

and thus,

$$P_3(x) - P_2(x) = \delta(f; x_0, x_1, x_2, x_3) \cdot (x - x_0)(x - x_1)(x - x_2) \tag{7.23}$$

Here we see how $P_3(x)$ is constructed by adding the appropriate term – the right-hand side of (7.23) – to $P_2(x)$. Next we substitute $x = x_3$ and recall that $P_3(x_3) = f(x_3)$. This leads to

$$f(x_3) - P_2(x_3) = \delta(f; x_0, x_1, x_2, x_3) \cdot (x_3 - x_0)(x_3 - x_1)(x_3 - x_2) \tag{7.24}$$

Because x_3 could have been any point in the interval $[a, b]$, we can rewrite (7.24) in the form

$$f(x) - P_2(x) = \delta(f; x_0, x_1, x_2, x) \cdot (x - x_0)(x - x_1)(x - x_2) \tag{7.25}$$

Because $[f(x) - P_2(x)]$ is precisely the remainder $R_2(x)$, we have

$$R_2(x) = f(x) - P_2(x) = \delta \cdot (x - x_0)(x - x_1)(x - x_2) \tag{7.26}$$

We emphasize once again that the relative positions of the distinct values x_0, x_1, x_2, and x_3 in $[a, b]$ are completely immaterial.

We remark that an expression of the form $(y_1 - y_0)/(x_1 - x_0)$, such as in (7.9), is termed a "divided difference," reflecting its nature. We can see that the expression for γ on the right-hand side of (7.13) is a divided difference of two divided differences and is therefore a "divided difference of second order." Similarly, the expression for δ in (7.22) is a "divided difference of third order." The laboratory participants should have no difficulty in defining divided differences of fourth and higher orders. Moreover, it can be shown that in general the remainder $R_n(x) = f(x) - P_n(x)$ is given by an appropriate divided difference of order $(n+1)$, multiplied by the product $(x - x_0)(x - x_1) \cdots (x - x_n)$. In this connection see Exercise 7.11.

Next, we use our knowledge of the structure of the remainder $R_n(x)$ to bound $R_n(x)$ by *error-bounding functions*.

7.5 Error-bounding functions

To estimate the quality of our approximations as expressed by (7.17) and (7.26), we first try to find bounds M_2 and M_3 such that for all x in $[a, b]$

$$|\gamma(f; x_0, x_1, x)| \leq M_2 \tag{7.27}$$

$$|\delta(f; x_0, x_1, x_2, x)| \leq M_3 \tag{7.28}$$

We use the notation M_2 and M_3 because γ and δ are divided differences of order 2 and 3, respectively. To compute $\gamma(f; x_0, x_1, x)$ and $\delta(f; x_0, x_1, x_2, x)$ for various values of x in $[a, b]$, we first have to choose the interpolation points (x_0, y_0), (x_1, y_1), and (x_2, y_2) . To make this choice, we must generate a sequence of x's for which the values of $y = f(x)$ are known, but we should not use the built-in library function for f because the construction of such a built-in function is the very purpose of the assignment. The collection of all the generated possible interpolation points will enable us to compute corresponding values of γ and δ and thus get an indication of the size of M_2 and M_3. We now demonstrate such computations for the function $\sqrt{x}$, even though it has been fully treated by iterations in Chapter 2, because of its simplicity and typical behavior.

Suppose we want to approximate $y = f(x) = \sqrt{x}$ in $[1, 4]$ (see Section 2.2 for the significance of the interval $[1, 4]$ in this example). A collection of possible interpolation points $(t_j, f(t_j))$ can, for example, consist of

$$t_j = \left(1 + \frac{j}{100}\right)^2, \qquad f(t_j) = 1 + \frac{j}{100}, \qquad j = 0, 1, \ldots, 100 \tag{7.29}$$

In other words, the values of $x = t_0, t_1, \ldots, t_{100}$ were chosen so as to have an exact square root. This collection, denoted by T, consists therefore of 101 points. We choose two or three interpolation points from T, to construct $P_1(x)$ or $P_2(x)$. In addition, let us denote by $\widetilde{T}$ the collection of all points in T except the points chosen as interpolation points. Now, M_2 and M_3 are computed as follows:

$$\begin{aligned} M_2 &\approx \operatorname*{Max}_{t_j \in \widetilde{T}} |\gamma(f; x_0, x_1, t_j)| \\ &= \operatorname*{Max}_{t_j \in \widetilde{T}} \left| \frac{\frac{f(t_j) - y_0}{t_j - x_0} - \frac{y_1 - y_0}{x_1 - x_0}}{t_j - x_1} \right| \end{aligned} \tag{7.30}$$

$$M_3 \approx \operatorname*{Max}_{t_j \in \widetilde{T}} |\delta(f; x_0, x_1, x_2, t_j)|$$

$$= \operatorname*{Max}_{t_j \in \widetilde{T}} \left| \frac{\frac{1}{t_j - x_1}\left(\frac{f(t_j) - y_0}{t_j - x_0} - \frac{y_1 - y_0}{x_1 - x_0}\right) - \frac{1}{x_2 - x_0}\left(\frac{y_2 - y_1}{x_2 - x_1} - \frac{y_1 - y_0}{x_1 - x_0}\right)}{t_j - x_2} \right| \tag{7.31}$$

In other words, we replace M_2 and M_3 by the maxima of $|\gamma|$ and $|\delta|$ in $\widetilde{T}$. It should be emphasized that the collection T (and thus $\widetilde{T}$) can be made as dense as we want by choosing $t_j = (1 + j/10^k)^2$, $j = 0, 1, \ldots, 10^k$, where $k > 2$. The denser the collection T, the better the estimates for M_2 and M_3. In this connection see Exercises 7.12 and 7.13. We emphasize again that the built-in library function has *not* been used; it will be activated *only* to verify, finally, the validity of the error bounds we find. Once M_2 and M_3 have been computed through (7.30) and (7.31), we have

$$|f(x) - P_1(x)| \leq M_2|(x - x_0)(x - x_1)| = E_2(x) \tag{7.32}$$

$$|f(x) - P_2(x)| \leq M_3|(x - x_0)(x - x_1)(x - x_2)| = E_3(x) \tag{7.33}$$

The error-bounding functions $E_2(x)$ and $E_3(x)$ are easy to compute and furnish an a priori estimate of the expected quality of approximation. Furthermore, they will guide us in the choice of "better" interpolation points from the collection, so as to reduce the maximal approximation error in $[a, b]$.

At this point we present, as an example, the results of a program that demonstrate the role of $E_3(x)$ as an error-bounding function. This program generates a table and a graph of $E_3(x)$, showing its relation to the graph of $|f(x) - P_2(x)|$. Specifically, we take $f(x) = \sqrt{x}$ in $[1, 4]$ and approximate it by $P_2(x)$ with the three initial interpolation points

$$(1.21, 1.1), \qquad (2.56, 1.6), \qquad (3.61, 1.9) \tag{7.34}$$

This triplet of points is one of many triplets of possible interpolation points we can choose from T for experimentation. Note that for the computation of $|\sqrt{x} - P_2(x)|$, the built-in $\sqrt{x}$ library function was used (for this purpose it represents the "genuine" square-root function). The built-in function has been used only to demonstrate the validity of the bounding properties of $E_3(x)$.

Even though we shall present mathematical arguments that lead to the choice of better interpolation points, it is worthwhile at this stage for the laboratory participants to experiment with various interpolation

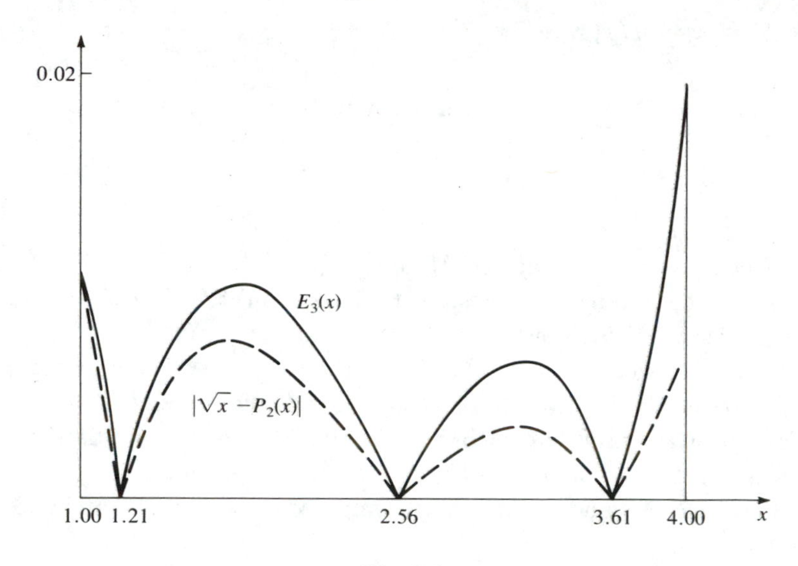

Fig. 7.2

points taken from T. In this connection see Exercise 7.14. When we ran the program that approximates $f(x) = \sqrt{x}$ in $[1, 4]$ by $P_2(x)$, based on the interpolation points (7.34), we obtained the graph in Fig. 7.2.

We see in Fig. 7.2 how $E_3(x)$ actually bounds the error $|\sqrt{x} - P_2(x)|$. This error was computed, as mentioned, using the library built-in $\sqrt{x}$ and is shown by the dotted line in the figure. The solid curve represents $E_3(x)$. In the left subinterval $1 \le x \le 1.21$, these two curves are so close that they are almost indistinguishable.

Now that we know the structure of the error-bounding functions $E_2(x)$, $E_3(x), \ldots, E_n(x)$, we want to find their maxima, so that we can pick the interpolation points judiciously and reduce the maximal errors. This subject is treated in the next section.

7.6 Bounds for the maximal errors

In this section we want to determine bounds for the error-bounding functions $E_2(x)$ and $E_3(x)$, defined in (7.32) and (7.33), where the constants M_2 and M_3 are assumed to have been computed by means of (7.30) and (7.31). Our understanding is (*for the time being*) that the pair of interpolation points x_0, x_1 in (7.32), as well as the triplet x_0, x_1, x_2 in (7.33), have been chosen and remain fixed.

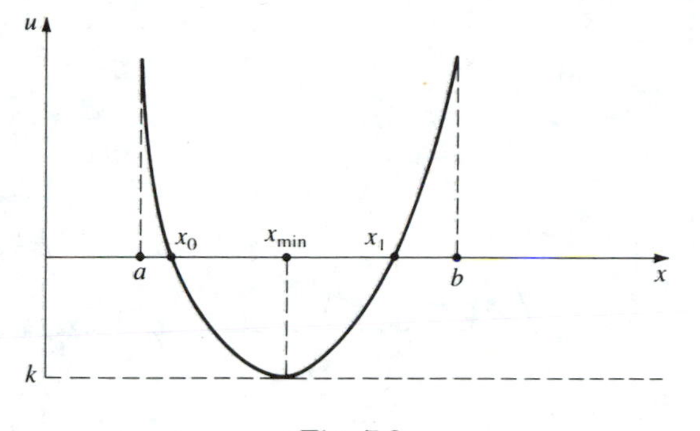

Fig. 7.3

To study the behavior of $E_2(x) = M_2|(x - x_0)(x - x_1)|$, we examine the parabola associated with it, given by

$$u = (x - x_0)(x - x_1) = x^2 - (x_0 + x_1)x + x_0x_1 \tag{7.35}$$

As preparation for also handling the cubic polynomial associated with $E_3(x)$, we find the minimum of the parabola (7.35) by requiring the existence of one (double) solution (see Fig. 7.3) to the quadratic equation

$$x^2 - (x_0 + x_1)x + x_0x_1 = k \tag{7.36}$$

Equation (7.36) has one (double) solution if and only if we have $(x_0 + x_1)^2 - 4(x_0x_1 - k) = 0$, that is,

$$k = x_0x_1 - (\frac{x_0 + x_1}{2})^2 = -(\frac{x_1 - x_0}{2})^2 = u_{\min} \tag{7.37}$$

Thus (see Fig. 7.3),

$$B_2 = \operatorname*{Max}_{a \leq x \leq b} E_2(x) = M_2 \cdot \operatorname{Max}\{u(a), -u_{\min}, u(b)\} \tag{7.38}$$

and we have found the constant bound of $E_2(x)$ and hence also of $|f(x) - P_1(x)|$. This number can be precomputed and therefore gives an a priori estimate of the expected error of approximation.

Analogously, because $E_3(x) = M_3|(x - x_0)(x - x_1)(x - x_2)|$, we are interested in the cubic

$$v = (x - x_0)(x - x_1)(x - x_2) = x^3 + px^2 + qx + r \tag{7.39}$$

where

$$\begin{aligned} p &= -(x_0 + x_1 + x_2) \\ q &= (x_0x_1 + x_1x_2 + x_2x_0) \\ r &= -x_0x_1x_2 \end{aligned} \tag{7.40}$$

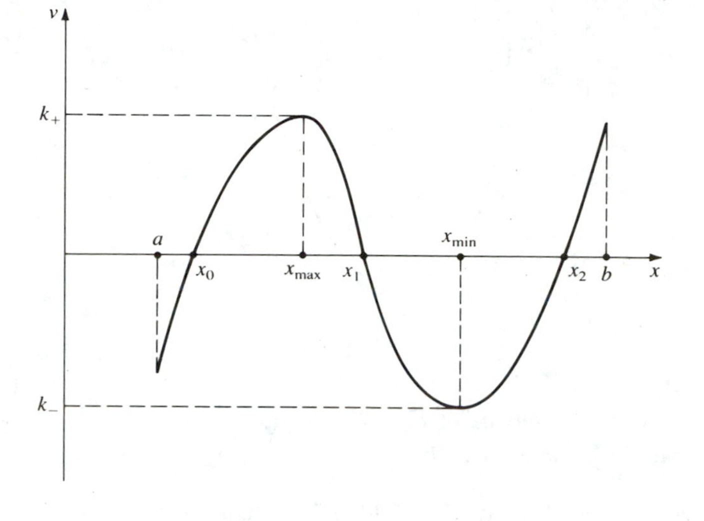

Fig. 7.4

We want to find the values k_+ and k_- of k, for which

$$x^3 + px^2 + qx + r = k \tag{7.41}$$

has a double root (see Fig. 7.4); that is, we want the values of k for which (7.41) is equivalent to

$$(x - \alpha)^2(x - \beta) = 0 \tag{7.42}$$

By comparing the coefficients in (7.41) and (7.42), we get α, β, and k in terms of p, q, and r. In particular,

$$k_\pm = \frac{27r + 2p^3 - 9pq}{27} \pm \frac{2(p^2 - 3q)^{3/2}}{27} \tag{7.43}$$

The laboratory participants will be asked to verify that $p^2 > 3q$ for distinct interpolation points. In this connection see Exercises 7.15 and 7.16.

Collecting our results, we find that the constant bounding $E_3(x)$, and thus also $|f(x) - P_2(x)|$, is given by

$$B_3 = \underset{a \le x \le b}{\mathrm{Max}}\ E_3(x) = M_3 \cdot \mathrm{Max}\{-v(a), v_{\max}, -v_{\min}, v(b)\} \tag{7.44}$$

where, clearly, $v_{\max} = k_+$ and $v_{\min} = k_-$, as can be seen in Fig. 7.4. The number B_3 constitutes an a priori estimate of the expected error of approximation when three interpolation points are used.

We now demonstrate the effectiveness of the previously mentioned

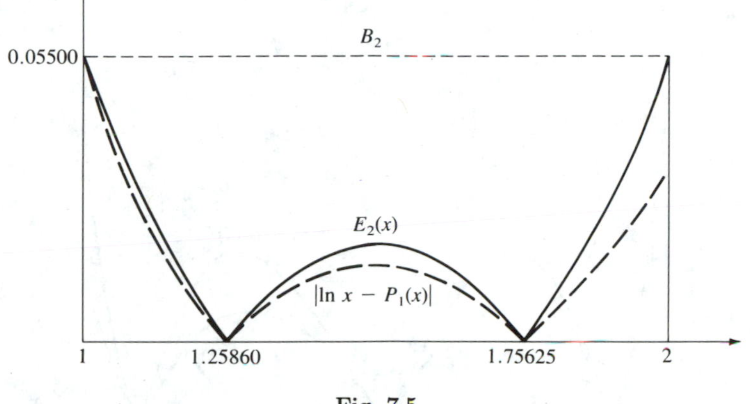

Fig. 7.5

bounds by using polynomial approximations of $f(x) = \ln x$ in $[1, 2]$. The interval $[1, 2]$ has been chosen because, for any positive t,

$$\ln t = \ln(x \cdot 2^m) = \ln x + m \ln 2 \tag{7.45}$$

where $1 \leq x < 2$, and $\ln 2$ is computed to a great accuracy once and for all (see Exercises 3.2 and 3.12). As our collection T of possible interpolation points, we take the points $(t_j, f(t_j))$, where

$$t_j = (2^{1/128})^j, \qquad f(t_j) = \frac{j}{128} \ln 2, \qquad j = 0, 1, \dots, 128 \tag{7.46}$$

The constant $2^{1/128}$ should be precomputed by seven consecutive square-root extractions (root extraction was a laboratory assignment in Chapter 2). From this collection of 129 points we choose two or three interpolation points at a time for the construction of our polynomial approximations in various numerical experiments. Figure 7.5 shows $|\ln x - P_1(x)|$, $E_2(x)$ and the constant bound B_2, for a typical linear approximation based on the two points $(1.24860, 0.22202)$ and $(1.75625, 0.56318)$, or $j = 41$ and 104 in (7.46). The computation that generated Fig. 7.5 also yielded $M_2 = 0.292309$, computed by (7.30), and the bounding constant $B_2 = 0.05496$.

In Fig. 7.6 we see $|\ln x - P_2(x)|$, $E_3(x)$, and B_3 based on the following three interpolation points: point 18 in T, $(1.10238, 0.09747)$; point 75, $(1.50101, 0.40614)$; and point 119, $(1.90486, 0.64441)$. In this computation we also found $M_3 = 0.138216$ and $B_3 = 0.00642$.

The laboratory participants should construct various polynomial approximations based on various interpolation points from T to gain an

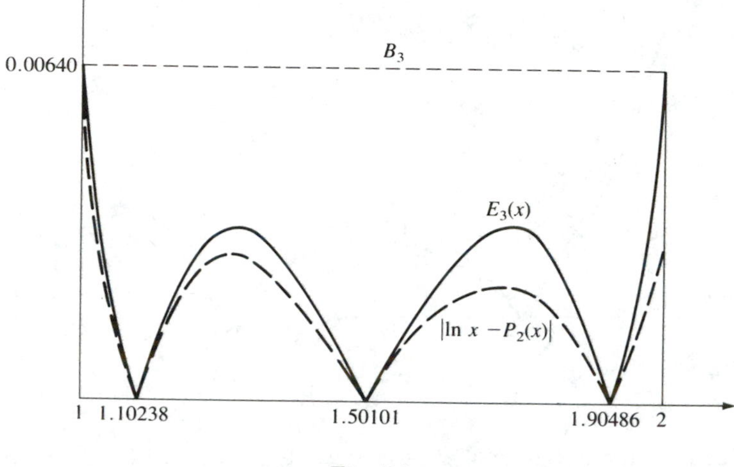

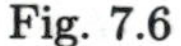
Fig. 7.6

intuitive feeling for the error behavior vs. the error-bounding function and bounding constant (see Exercises 7.17 and 7.18). It is especially noteworthy that the largest values of the error and the error-bounding functions in Figs. 7.5 and 7.6 are located at the interval endpoints. We could, of course, have taken these endpoints as interpolation points, in which case the error there would have vanished, but at the cost of a considerable increase inside the interval. It is suggested (see Exercises 7.19 and 7.20) that participants try such an approximation as one of the numerical experiments. This being the situation, it is natural to inquire whether a compromise between these two extreme error behaviors can be achieved. In other words, how can the interpolation points be chosen so as to reduce the *overall error bound*, thus improving the approximation. This should not be done by trial and error, using the library built-in function f, but by an analytical examination of the error-bounding functions $E_2(x)$ and $E_3(x)$. This ought to lead us to reduced error bounds B_2 and B_3, and thus provide us with improved estimates of the maximal errors expected. Clearly we should not be using an available built-in function f in the very process of constructing our own built-in function.

7.7 Reduction of the error bounds

The strategy for improving the error distribution in the interval is to impose a "balance" among the maxima of the error-bounding function.

For three interpolation points, this means (see Fig. 7.4) that we require

$$-v(a) = v_{\text{max}} = -v_{\text{min}} = v(b) \tag{7.47}$$

If such a balance is achieved, the error-bounding function is said to possess the *equal ripple property* in $[a, b]$. We note that the number of such ripples exceeds the number of interpolation points by one. To simplify the analysis leading to equal ripples, we take the interval to be $[-1, 1]$. This does not limit the generality because we can easily transform the interval $[a, b]$ to $[-1, 1]$ by the linear transformation

$$x_{\text{new}} = \frac{2}{b-a}x_{\text{old}} - \frac{b+a}{b-a} \tag{7.48}$$

and finally transform all the results back to $[a, b]$. In this connection see Exercise 7.21. Thus, let us examine (7.39), the cubic $v = (x - x_0)(x - x_1)(x - x_2)$ in the interval $[-1, 1]$. It stands to reason to choose the interval midpoint as one of the interpolation points, say $x_1 = 0$, and thus $r = -x_0x_1x_2 = 0$. Imposing the condition $-v(-1) = v(1)$, we find that $p = 0$ too, and then by (7.43) that $-v_{\text{min}} = -k_- = k_+ = v_{\text{max}}$. Consequently,

$$v = (x - x_0)(x - 0)(x - x_2) = x^3 + qx = x(x + \sqrt{-q})(x - \sqrt{-q}) \tag{7.49}$$

and hence

$$-x_0 = x_2 = \sqrt{-q} \qquad (\text{note}: \ q = x_0x_2 = -x_2^2 < 0) \tag{7.50}$$

To make all four ripples equal, we must satisfy just one additional condition, $v_{\text{max}} = v(1)$. This takes the form

$$-\frac{2}{9}q\sqrt{-3q} = 1 + q \tag{7.51}$$

Using the notation $z = \sqrt{-3q}$, (7.51) can be rewritten as

$$2z^3 + 9z^2 - 27 = 0 \tag{7.52}$$

which is equivalent to

$$(2z - 3)(z + 3)^2 = 0 \tag{7.53}$$

The value of z we are looking for has to be positive, so the relevant solution of (7.53) is $z = \frac{3}{2}$ (i.e., $q = -\frac{3}{4}$). Thus, (7.50) yields $-x_0 = x_2 = \sqrt{3}/2$, and (7.49) becomes $v = x(x^2 - \frac{3}{4})$. This, finally, is the cubic with equal ripples in $[-1, 1]$, and we find the size of those ripples to be $-v(-1) = v_{\text{max}} = -v_{\text{min}} = v(1) = \frac{1}{4}$. The laboratory participants will confirm (see Exercise 7.22) that on transforming $[-1, 1]$ back to

$[a, b]$, the special interpolation points in $[a, b]$ that correspond to $x_0 = -\sqrt{3}/2, x_1 = 0$, and $x_2 = \sqrt{3}/2$ in $[-1, 1]$ are

$$\begin{aligned} x_0 &= \frac{b+a}{2} - \left(\frac{b-a}{2}\right)\frac{\sqrt{3}}{2} \\ x_1 &= \frac{b+a}{2} \\ x_2 &= \frac{b+a}{2} + \left(\frac{b-a}{2}\right)\frac{\sqrt{3}}{2} \end{aligned} \tag{7.54}$$

and the common size of the ripples is given by

$$\text{ripple size} = \frac{1}{4}\left(\frac{b-a}{2}\right)^3 \tag{7.55}$$

We now apply all of these results to $f(x) = \ln x$ in $[1, 2]$. The special interpolation points (7.54) in this interval are

$$x_0 = 1.0669873, \qquad x_1 = 1.5, \qquad x_2 = 1.9330127 \tag{7.56}$$

However, these points are not included in T, the prepared collection (7.46) of possible interpolation points. We therefore take the points from T that are closest to the special points (7.56). These turn out to be the points $(t_j, f(t_j))$ in (7.46) with $j = 12,\ 75$, and 122, that is,

$$\begin{aligned} (x_0, y_0) &= (1.06714, 0.06498) \\ (x_1, y_1) &= (1.50101, 0.40614) \\ (x_2, y_2) &= (1.93606, 0.66066) \end{aligned} \tag{7.57}$$

Because the abscissas of these points are not identical to (7.56), we expect to get an error-bounding function with almost equal ripples. We could have chosen points even closer to (7.56) by preparing a denser collection T, by replacing 128 with 256, say, in (7.46) and generating $(256+1)$ possible interpolation points. Further refinements are possible, of course, at will.

After running our program with the interpolation points (7.57), we produced the graph in Fig. 7.7 with almost equal ripples, as expected. We also obtained the numerical results $M_3 = 0.139655$ and $B_3 = 0.0044$. In connection with the refinement of the collection T of possible interpolation points, see Exercises 7.23, 7.24, and 7.25.

Comparison of the graph in Fig. 7.7 with that in Fig. 7.6 (obtained before the improvement) shows that as a result of approaching the equal ripple property of $E_3(x)$, the error bound B_3 decreased by a factor of about $\frac{2}{3}$. This improvement is accompanied by a considerable reduction

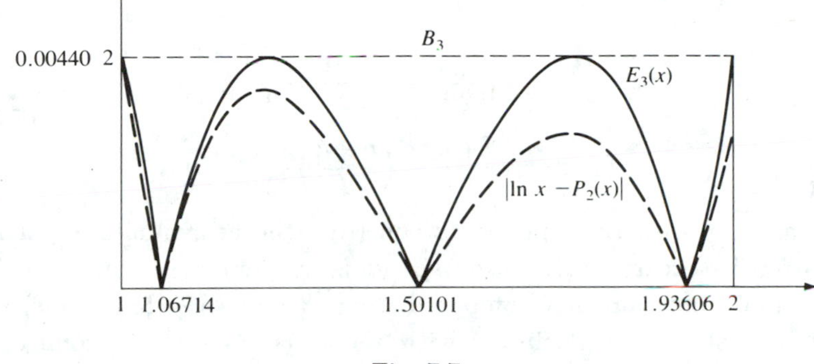

Fig. 7.7

of the maximal approximation error $|\ln x - P_2(x)|$. Note that this approximation error was computed merely as a check, which is why we took the liberty of using the logarithmic library function for that purpose. The interpolation points that led to Fig. 7.6 turned out to be a fairly good choice to begin with, and therefore the improvement factor achieved (reflected in Fig. 7.7) was just about $\frac{2}{3}$. Had we chosen equally spaced initial interpolation points (which would have been the natural thing to do), the improvement factor would have been far more dramatic. In this connection see Exercises 7.26, 7.27, and 7.28. Our main conclusion in this section is that the suggested mathematical analysis leading to the improved interpolation points (7.54) guides us in choosing judicious points, through which the overall error bound is reduced considerably. The resulting approximation polynomial $P_2(x)$ is hence known as the *near-minimax second-degree polynomial* approximating $f(x)$ in $[a, b]$. Moreover, the analysis also supplies us with an a priori estimate of the quality of the approximation.

It should be noted that the improved interpolation points (7.54), which resulted from the error analysis presented in this section, are independent of the function $f(x)$ that we want to approximate. The nature of the particular function $f(x)$ is reflected only in the corresponding size of M_3. This is so because in our analysis we considered M_3 and $\text{Max}|(x - x_0)(x - x_1)(x - x_2)|$ separately. Otherwise, a much deeper analysis (beyond our scope here) would have been necessary, leading to the minimax rather than near-mimimax polynomial.

At this point, laboratory participants should carry out the entire analysis, this time for the case of two interpolation points, based on (7.38). They will find the two improved interpolation points (see Exercise 7.29) to be

$$\begin{aligned} x_0 &= \frac{b+a}{2} - \left(\frac{b-a}{2}\right)\frac{\sqrt{2}}{2} \\ x_1 &= \frac{b+a}{2} + \left(\frac{b-a}{2}\right)\frac{\sqrt{2}}{2} \end{aligned} \tag{7.58}$$

We actually used these points to construct the near-minimax polynomial of first degree $P_1(x)$, corresponding to $f(x) = \ln x$ in $[1, 2]$. We then ran a program that computed M_2 and $B_2 = M_2 \cdot \mathrm{Max}\, E_2(x)$ and obtained almost equal ripples with $B_2 = 0.0374$. In this connection see Exercise 7.30. We recall that the overall error bound B_3, based on three interpolation points and corresponding to an almost equal rippled $E_3(x)$, was found to be 0.0044 for this case (see Fig. 7.7). It follows that the overall error bound is reduced by a factor of 8.5 when three rather than two interpolation points are used. This gives an indication of what further improvements in the quality of approximation could be achieved if even more interpolation points were used. Indeed, it turns out that had we used seven interpolation points for this case (generalizing our ideas here), we would have obtained an overall error bound of about 10^{-6}. The construction of computer library functions is just a natural extension of these findings, and we shall dwell upon it in detail in Chapter 8.

The generalization of our analysis, designed to produce an equal rippled $E_n(x) = M_n \cdot \mathrm{Max}|(x-x_0)(x-x_1)\cdots(x-x_{n-1})|$, was found by P. Chebyshev. He showed that for n interpolation points $x_0, x_1, x_2, \ldots,$ x_{n-1}, we have

$$x_k = \frac{b+a}{2} - \left(\frac{b-a}{2}\right) \cdot \cos\left[\left(k + \frac{1}{2}\right)\frac{\pi}{n}\right] \tag{7.59}$$

where $k = 0, 1, 2, \ldots, n-1$. These points, known as *Chebyshev points*, depend exclusively upon the number n and the underlying interval $[a, b]$. We verify that (7.59) reduces to (7.58) and (7.54) for $n = 2$ and $n = 3$, respectively. The Chebyshev points for $n = 12$ are shown in Fig. 7.8, where $\alpha = \pi/12$. They are far from being equidistant but are obviously symmetric about the interval's midpoint and crowd together toward its endpoints (which are not included).

The subject of interpolative approximations has been illustrated here

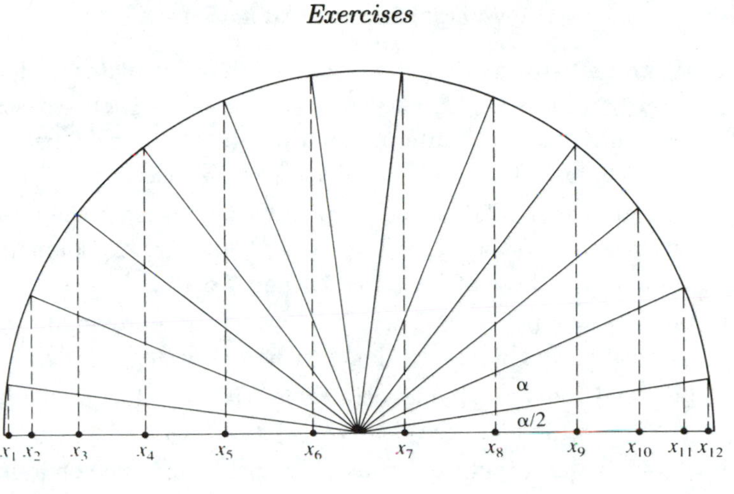

Fig. 7.8

mainly through the logarithmic function. Other important and intriguing examples are the trigonometric functions, whose approximation will be discussed at the beginning of the next chapter.

Exercises

7.1 Express the values of $\sin 7.5°$, $\cos 7.5°$, $\sin 82.5°$, $\sin 22.5°$, and $\cos 3.75°$ in terms of appropriate square roots. Note that the actual computation of square roots was discussed in Chapter 2.

7.2 Prove that through three distinct given points (not collinear) there passes a unique parabola. Assume that two parabolas can be passed through the three points, and arrive at a contradiction.

7.3 Prove the uniqueness of the nth-degree polynomial that passes through $(n+1)$ distinct points. Assume two such polynomials, examine their difference, and deduce a contradiction.

7.4 Write the Lagrange interpolation polynomial for $n = 3$. Then express a_0, a_1, a_2, and a_3 in terms of the coordinates of the four interpolation points, so that you can write $P_3(x) = a_3x^3 + a_2x^2 + a_1x + a_0$.

7.5 Write and run a computer program that calculates $R_6(x) = \sin x - P_6(x)$, for x increasing from $x = 0$ to $x = \pi/2$, using increments of $\pi/1440$. Verify that $\text{Max}\,|R_6(x)| < 1.3 \cdot 10^{-6}$. We recommend that $P_6(x)$ be introduced into the program as a separate procedure based on Lagrange's interpolation formula with the interpolation points (7.1).

7.6 Write and run a computer program that calculates and plots $R_3(x) = \sin x - P_3(x)$, $0 \leq x \leq \pi/2$, where $P_3(x)$ is based on the four equidistant interpolation points $x = 0$, $\pi/6$, $\pi/3$, and $\pi/2$. Compare the results with those of Exercise 7.5.

7.7 Repeat Exercise 7.6, this time with $P_4(x)$ based on the interpolation points $x = \pi/12$, $\pi/6$, $\pi/4$, $\pi/3$, and $5\pi/12$. Compare the results with those of Exercises 7.5 and 7.6.

7.8 Given $y = f(x) = 1/x$, which we want to approximate in the interval $[1, 4]$. Carry out the following:

(a) Find the straight line coinciding with $y = 1/x$ at the interval's endpoints, that is, at $(1, 1)$ and $(4, \frac{1}{4})$.

(b) Find the straight line that is parallel to the line you found in (a) and tangent to $y = 1/x$. (Use an appropriate quadratic equation whose two roots will be identical.)

(c) Show that the line you found in (b) is tangent to $y = 1/x$ at $x = 2$.

(d) Find the straight line, to be denoted by $P_1(x)$, parallel to and midway between the lines you found in (a) and (b).

(e) Set $R(x) = 1/x - P_1(x)$, and show that $R(1) = -R(2) = R(4)$, that is, that $P_1(x)$ possesses the equal ripple property with respect to $1/x$ in $[1, 4]$.

(f) Draw the graph of $R(x) = 1/x - P_1(x)$ in $[1, 4]$, representing the error incurred by approximating $1/x$ by $P_1(x)$. Verify that this error is smeared uniformly, in the equal ripple sense.

(g) Find the points at which $1/x = P_1(x)$, that is, the interpolation points of $P_1(x)$.

(h) Verify that the line $P_1(x) = x/3 + 17/24$ constructed in Section 2.5, which approximates $y = \sqrt{x}$ in $[1, 4]$, actually possesses the equal ripple property with respect to $\sqrt{x}$. Draw the graph of $R(x) = \sqrt{x} - (x/3 + 17/24)$ in $[1, 4]$ so you can see the equal ripples.

7.9 Given $P_2(x) = P_1(x) + \gamma \cdot (x - x_0)(x - x_1)$, where $P_1(x)$ is defined in (7.9) and hence is the line passing through (x_0, y_0) and (x_1, y_1), as does $P_2(x)$. Show that if $P_2(x)$ is to pass also through (x_2, y_2), then

$$\gamma = \frac{\dfrac{y_2 - y_0}{x_2 - x_0} - \dfrac{y_1 - y_0}{x_1 - x_0}}{x_2 - x_1}$$

Show also that γ can be expressed equivalently in the form

$$\gamma = \frac{\dfrac{y_2 - y_1}{x_2 - x_1} - \dfrac{y_1 - y_0}{x_1 - x_0}}{x_2 - x_0}$$

obtained by interchanging the points (x_0, y_0) and (x_1, y_1). Other equivalent forms can be obtained by interchanging any pair of the interpolation points. This is to be expected because the entire analysis made no use of the relative positions of the interpolation points.

7.10 Show that Newton's form (7.18) and Lagrange's form (7.19) are equivalent forms of the same interpolation parabola.

7.11 Set $R_3(x) = f(x) - P_3(x) = \lambda \cdot (x - x_0)(x - x_1)(x - x_2)(x - x_3)$, and show that λ is the appropriate divided difference of fourth order.

7.12 Write and run a computer program that finds maximum values as defined in (7.30) and (7.31). Take $t_j = (1 + j/10^k)^2$ and $f(t_j) = \sqrt{t_j} = (1 + j/10^k)$, for $j = 0, 1, 2, \ldots, 10^k$, and carry out the computations for $k = 2$, $k = 3$, and $k = 4$. Take $x_0 = 1$ and $x_1 = 4$ in (7.30), and add a third point $x_2 = 1.5^2 = 2.25$ for (7.31). Note that the computations with the denser $\widetilde{T}$ (i.e., $k = 3$) yield higher values for the estimated M_2 and M_3. This is to be expected because the maximum taken over an enlarged collection $\widetilde{T}$ can only increase.

7.13 Repeat Exercise 7.12 for the function $f(x) = \sqrt[3]{x}$ in $[1, 8]$. This time take $t_j = (1 + j/10^k)^3$ and $f(t_j) = \sqrt[3]{t_j} = (1 + j/10^k)$ for $j = 0, 1, 2, \ldots, 10^k$, first for $k = 2$, then for $k = 3$, and finally for $k = 4$.

7.14 Write and run a computer program that outputs the graphs of $E_3(x)$ as well as $|\sqrt{x} - P_2(x)|$ in $[1, 4]$, similar to Fig. 7.2, where $P_2(x)$ is based on $x_0 = 1.2^2 = 1.44$, $x_1 = 1.5^2 = 2.25$, and $x_2 = 1.8^2 = 3.24$. Repeat the entire process, this time using $x_0 = 1.00$ and $x_1 = 2.25$, and $x_2 = 4.00$. Compare the values of $\operatorname{Max} E_3(x)$, $1 \le x \le 4$ for these two cases and the case depicted in Fig. 7.2.

7.15 By comparing the coefficients in (7.41) and (7.42), find α and β in terms of r, p, and q. In addition, verify the expressions for $k_+ = v_{\max}$ and $k_- = v_{\min}$ as given in (7.43).

7.16 Prove that, for distinct interpolation points, $(p^2 - 3q)$ is positive, so that k_+ and k_- in (7.43) are real. [Hint: $a^2 + b^2 + c^2 = \frac{1}{2}(a^2 + b^2) + \frac{1}{2}(b^2 + c^2) + \frac{1}{2}(c^2 + a^2)$.]

7.17 For $f(x) = \ln x$, $1 \leq x \leq 2$, choose *two* interpolation points from the collection T described in (7.46), then write a program that plots $E_2(x)$ as well as $|\ln x - P_1(x)|$ in this interval. Also print the values of M_2 and B_2, bearing in mind that in computing the values of $|\ln x - P_1(x)|$ you are using the computer's built-in logarithmic function. Carry out the preceding for various pairs of interpolation points, one of which being the pair corresponding to $j = 41$ and $j = 104$ in T so that you reproduce Fig. 7.5.

7.18 Repeat Exercise 7.17, this time using *three* interpolation points. As one of the triplets, take the points corresponding to $j = 18$, $j = 75$, and $j = 119$ in T, so that you reproduce Fig. 7.6.

7.19 Repeat Exercise 7.17 with the *pair* of interpolation points corresponding to $j = 0$ and $j = 128$ in T, that is, the endpoints of the interval $[1, 2]$. Compare the resulting overall bound B_2 with those you obtained in Exercise 7.17.

7.20 Repeat Exercise 7.18 with the *triplet* of interpolation points corresponding to $j = 0$, $j = 75$, and $j = 128$. Observe that the first and third points correspond to the interval's endpoints and that the second point is approximately the interval's midpoint. Compare the resulting overall bound B_3 with those you obtained in Exercise 7.18.

7.21 Consider the linear transformation (7.48)

$$x_{\text{new}} = \frac{2}{b-a} x_{\text{old}} - \frac{b+a}{b-a}$$

For $x_{\text{old}} = a$, $(b+a)/2$, and b, find the corresponding x_{new}. What is the inverse transformation carrying the interval $[-1, 1]$ back to $[a, b]$. Rewrite the original transformation in the form

$$x_{\text{new}} = \frac{2}{b-a}\left(x_{\text{old}} - \frac{b+a}{2}\right)$$

and show that it can be interpreted as a displacement followed by a contraction (or stretching).

7.22 Show that if the interval $[-1, 1]$ is transformed back to $[a, b]$, as shown in Exercise 7.21, then the special interpolation points $-\sqrt{3}/2$, 0, $\sqrt{3}/2$ are carried to the points given in (7.54). Also, verify that the equal ripples that were of common size $\frac{1}{4}$ in $[-1, 1]$, are now of size $\frac{1}{4} \cdot [(b-a)/2]^3$ in $[a, b]$.

7.23 Write and run a computer program that reproduces the results shown in Fig. 7.7 by means of the interpolation points (7.57). Note, especially, the almost equal ripples.

7.24 Prepare a collection T that is denser than (7.46) by replacing 128 with 256. From this new collection, select three interpolation points that are closest to those given in (7.56), and compare with (7.57).

7.25 Repeat Exercise 7.23 using the three interpolation points you found in Exercise 7.24. Note that the almost equal ripples found in Exercise 7.23 gave way to even better ones.

7.26 Repeat Exercise 7.23, using three interpolation points from the collection T that are closest to the values 1.25, 1.50, and 1.75. Compare the results with those obtained in Exercise 7.23. What can be said of the relative sizes of the overall error bounds obtained here and in Exercise 7.23.

7.27 Write and run a program that plots $E_3(x)$, corresponding to $f(x) = \sqrt{x}$ in $[1, 4]$. Use various triplets of points, at will, out of the collection T in (7.29). For completeness, also plot the graph of $|\sqrt{x} - P_2(x)|$.

7.28 Repeat Exercise 7.27, this time using three interpolation points from the collection T in (7.29) that are closest to the values in (7.54). Note the improved error behavior.

7.29 On the basis of Fig. 7.3 and equation (7.38), show that the two interpolation points, yielding an equal rippled $E_2(x)$, are given by (7.58). These points are the two-point analogue of the three-point case described in (7.54).

7.30 Write and run a program that plots $E_2(x)$ for $f(x) = \ln x$ in $[1, 2]$, using two interpolation points from (7.46) that are closest to the points you found in Exercise 7.29. Compare the results with those obtained in Exercise 7.17, and observe the improved error behavior.

8
Computer library functions

8.1 Built-in functions

The term *computer library functions* refers to the collection of built-in functions – $\sin x$, $\ln x$, e^x, $\arctan x$, to name but a few – that were installed in the computer's permanent memory. These built-in functions, of course, are efficient approximations of the abstract mathematical entities they represent. By efficiency we mean that every evaluation is performed with utmost speed and yields all the correct significant figures that are available on the computing device used. The construction of such built-in functions usually entails lengthy, computationally expensive preparations, which, however, are carried out only *once*. The first preparatory step is to reduce, as much as possible, the interval $[a, b]$ in which the given $f(x)$ is to be approximated (examples of this strategy can be found in Sections 2.2 and 7.6). To guarantee the desired correct significant figures, we also must control the *relative error* in the approximation of $f(x)$ in $[a, b]$. This issue will be discussed in Section 8.3.

We shall concentrate on *polynomial* approximations, making use of the results obtained in Chapter 7, and shall consider the possibility of constructing *rational* approximations in Sections 8.5 and 8.6. We start with polynomial approximation of trigonometric functions in the same spirit with which we treated $\ln x$ in Chapter 7.

8.2 Trigonometric library functions

We observe first that because the function $\cos x$ is periodic with period 2π, and thus $\cos(-x) = \cos x$, $\cos(\pi - x) = -\cos x$, $\cos(\pi + x) = -\cos x$, and $\cos(2\pi - x) = \cos x$, it is sufficient to approximate $\cos x$ in the first quadrant, $0 \leq x \leq \pi/2$ (see Exercise 8.1). Moreover, because $\sin x = \cos(\pi/2 - x)$, the approximation of the cosine function yields an approximation for the sine function as well.

Next, we want to build up a collection T of points in $[0, \pi/2]$, whose cosines are computable using trigonometric formulas, from which we can choose suitable interpolation points. To this end, we start by generating a very precise value for $\cos t$, for a small positive t, say,

$$t = \frac{1}{128} \cdot \frac{\pi}{2} \tag{8.1}$$

We chose 128 for definiteness, but any other power of 2, such as 256 or 512, could have been chosen. We recall the half-angle formula

$$\cos\frac{\alpha}{2} = \sqrt{\frac{1+\cos\alpha}{2}} \tag{8.2}$$

and apply it successively seven times, starting with $\alpha = \pi/2$, to obtain $\cos t$ for $t = \frac{1}{128}(\pi/2)$. Next we recall the trigonometric identity

$$\cos\beta + \cos\gamma = 2\cos\frac{\beta-\gamma}{2}\cos\frac{\beta+\gamma}{2} \tag{8.3}$$

and choose $\beta = (k+1)t$ and $\gamma = (k-1)t$. Thus, we reach

$$\cos\big[(k+1)t\big] = 2\cos t\cos kt - \cos\big[(k-1)t\big] \tag{8.4}$$

which we use recursively to get the values of $\cos 2t, \cos 3t, \ldots, \cos 128t$. We start this computation with $k = 1$ and use the value of $\cos t$ just obtained to find $\cos 2t$, and so on. All these calculations, including the calculation of $\cos t$, should be carried out with *double precision* to avoid numerical contamination by round-off errors (see Exercise 8.2). Although these preparatory computations are tedious, we emphasize that they are carried out *only once*, and they supply us with a collection T of 128 possible interpolation points. We stress that we can replace 128 by 256 or any other higher power of 2 and obtain as dense a collection T as we please. This rich collection of interpolation points will enable us to improve the approximation of $\sin x$, discussed in Section 7.3. This can be achieved by choosing better interpolation points, in the sense of reducing the relatively large approximation errors near the endpoints, depicted in Fig. 7.1.

Because our intention is to construct computer library functions, we must exercise care to ensure the desired number of correct significant figures. Accordingly, it is the relative error – not just the absolute error – that must be made sufficiently small.

8.3 Control of relative errors

When we want to construct a built-in approximating polynomial for a computer's library, to represent a function $f(x)$, we must take special care near points at which $f(x)$ vanishes. Consider, for example, the function $\sin x$ near $x = 0$. Not only do we want the built-in function to vanish at $x = 0$, but at the same time we must ensure a small relative error for small values of x. The reason is that we want to be sure of a preassigned number of correct significant figures. Suppose, for example, that the built-in function supplies the value $\sin x = 0.0000318$ for a given small x, and we also know that the last (and only the last) digit may be incorrect. In that case, though the error is less than 10^{-6}, we would only have two correct significant figures, and the relative error is of the size of a few hundredths – far more than 10^{-6}. The laboratory instructor should dwell on the importance of the relative-error concept. This can easily be done using, say, examples of errors in measurements. An error of one inch is certainly negligible when measuring the distance to the moon, but the same error will be disastrous when fitting a hat to a customer's head.

To ensure that our built-in function representing $\sin x$ carries a sufficiently small relative error, we proceed as follows. It is well known (from high-school studies of optics, motion of a pendulum, etc.) that for small values of x (in radians), $\sin x$ is very nearly equal to x. Accordingly, we require a similar behavior of the built-in function, in addition to its vanishing at $x = 0$. This can be achieved by constructing an approximating polynomial $P_n(x)$ for the function $g(x) = (\sin x)/x$ (rather than for $f(x) = \sin x$) and demanding that $P_n(0) = 1$. Multiplying this polynomial by x gives us an approximation for $\sin x$ in the form

$$xP_n(x) = x + (\text{higher order powers of } x) \tag{8.5}$$

which has the correct behavior near $x = 0$. The reason we chose to start by approximating $(\sin x)/x$ is that this function is nowhere near zero (see Exercise 8.3) in the underlying interval $[0, \pi/2]$ – in fact, it actually decreases (see Exercise 8.4) from unity to $2/\pi \approx 0.6366$. Thus, we have

$$\frac{\sin x}{x} = P_n(x) + \text{ error} \tag{8.6}$$

where the error is guaranteed (by construction) to be sufficiently small. It follows now that

$$\sin x = xP_n(x) + x \cdot (\text{error}) \tag{8.7}$$

where the new error term, $x\cdot$ (error), is sufficiently small relative to the value of $xP_n(x)$, which the computer's library supplies as $\sin x$. For example, if the built-in function supplies the value $0.743561 \cdot 10^{-8}$, say, and the construction was designed to guarantee a relative error less than 10^{-6}, then we know that all the displayed figures are correct.

To see how to handle the relative-error control near the roots of a function $f(x)$, not necessarily $\sin x$, we take another look at the preceding considerations. A little reflection shows that by approximating $g(x) = (\sin x)/x$, rather than $f(x) = \sin x$, we have effectively (temporarily) removed the root from $f(x)$. Suppose, then, that we have constructed an approximating polynomial $P_n(x)$ such that for every x in $[0, \pi/2]$ we have

$$\left| \frac{\sin x}{x} - P_n(x) \right| < \alpha \cdot 10^{-k} \tag{8.8}$$

where k expresses the required accuracy, and the value of α is to be determined. Multiplying both sides of (8.8) by the positive quotient $x/\sin x$, we obtain

$$\left| \frac{\sin x - xP_n(x)}{\sin x} \right| < \frac{x}{\sin x} \cdot \alpha \cdot 10^{-k} \tag{8.9}$$

Because the polynomial approximating $\sin x$ is precisely $xP_n(x)$, it follows that the left-hand side of (8.9) is the relative error in this approximation. Now, the quotient $x/\sin x$ on the right-hand side of (8.9) increases from the value 1 at $x = 0$ to $\pi/2$ at $x = \pi/2$ (see again Exercises 8.3 and 8.4), and so we have

$$\left| \frac{\sin x - xP_n(x)}{\sin x} \right| < \alpha\pi \cdot \frac{1}{2} \cdot 10^{-k} \tag{8.10}$$

Finally, we see that if we choose $\alpha \leq 1/\pi$, say $\alpha = 0.31$, the relative error is controlled so that k correct significant figures are assured in the approximation of $\sin x$ by $xP_n(x)$.

As a second example, we consider $f(x) = \ln x$ in the interval $[1, 2]$, which suffices for the computation of the logarithm of any positive number [see equation (7.45) and the following sentence]. Because $\ln 1 = 0$, we examine $g(x) = \ln x/(x-1)$ in this interval. As seen in Exercises 8.5 and 8.6, the function $g(x) = \ln x/(x-1)$ decreases from unity at $x = 1$ to $\ln 2 \approx 0.6931$ at $x = 2$, and thus we approximate $g(x)$ rather than $f(x) = \ln x$. In this way we have again "removed" the root from $f(x)$. Suppose, then, that we have constructed an approximating polynomial $P_n(x)$ such that for every x in $[1, 2]$ we have

$$\left| \frac{\ln x}{x-1} - P_n(x) \right| < \beta \cdot 10^{-k} \tag{8.11}$$

where β is to be determined. Multiplying both sides by the positive quotient $(x-1)/\ln x$, we obtain

$$\left|\frac{\ln x-(x-1)P_n(x)}{\ln x}\right| < \frac{x-1}{\ln x}\cdot\beta\cdot 10^{-k} \tag{8.12}$$

As before, the left-hand side of (8.12) is the relative error in the approximation of $\ln x$ by $(x-1)P_n(x)$. By Exercise 8.6 we have $1 \le (x-1)/\ln x \le 1/\ln 2$, and thus,

$$\left|\frac{\ln x-(x-1)P_n(x)}{\ln x}\right| < \frac{2\beta}{\ln 2}\cdot\frac{1}{2}\cdot 10^{-k} \tag{8.13}$$

We see that if we choose $\beta \le \frac{1}{2}\ln 2$, say $\beta = 0.34$, the relative error is controlled, and so k significant figures are assured in the approximation of $\ln x$ by $(x-1)P_n(x)$.

More generally, if the function $f(x)$ to be approximated has a root at $x=r$ and if the function $g(x)=f(x)/(x-r)$ does not vanish at $x=r$, then we construct $P_n(x)$ to approximate $g(x)$ rather than $f(x)$. Under these circumstances $f(x)$ is said to have a *simple root* at $x=r$. Next, we examine the quotient $(x-r)/f(x)$ and try to find bounds m and M such that

$$0 < m \le \left|\frac{x-r}{f(x)}\right| \le M \tag{8.14}$$

for all x in the relevant interval. Having found these bounds, we see that

$$\left|\frac{f(x)}{x-r}-P_n(x)\right| < \gamma\cdot 10^{-k} \tag{8.15}$$

implies

$$\left|\frac{f(x)-(x-r)P_n(x)}{f(x)}\right| < 2\gamma M\cdot\frac{1}{2}\cdot 10^{-k} \tag{8.16}$$

which yields the desired relative-error control for $\gamma \le \frac{1}{2}M$.

If the root $x=r$ of $f(x)$ is not simple but of multiplicity $p \ge 2$, that is,

$$\frac{f(x)}{(x-r)^p} \ne 0 \qquad \text{for} \quad x=r \tag{8.17}$$

but

$$\frac{f(x)}{(x-r)^q} = 0, \quad \text{for} \quad x=r,\ q=0,1,2,\ldots,p-1 \tag{8.18}$$

then the entire procedure should be carried out using the function $g(x) = f(x)/(x-r)^p$. In this connection see Exercises 8.7, 8.8, and 8.9.

8.4 Computational complexity

Having discussed the construction of the approximating polynomial so as to control the relative error, we now study the form into which the approximating polynomial $P_n(x)$ should be cast for permanent storage in the computer's library, so that its evaluation (whenever called on) will be performed efficiently.

Suppose the polynomial has already been constructed and put into the form

$$P_n(x) = a_n x^n + a_{n-1}x^{n-1} + \cdots + a_1 x + a_0 \tag{8.19}$$

Obviously (see Chapter 7), a considerable amount of computational work is involved in calculating $a_0, a_1, \ldots, a_n$. But this work is done only once, after which the polynomial (8.19) is ready for use and re-use a vast number of times. Our next objective is to show that the form (8.19) can be improved in a way that renders the computations of this library function more efficient.

Let the laboratory participants count the number of multiplications and additions needed to evaluate $P_n(x)$ in (8.19) for a given x. They will find that n additions and $(2n-1)$ multiplications are required. Let us define the computational work involved in executing a single addition to be "one computing unit." We assume that a single multiplication requires about one computing unit (this is indeed the case now). Accordingly, one evaluation of (8.19) consumes roughly $3n$ computing units.

To increase the computational efficiency, let us rewrite (8.19) in the so-called Horner's form (see Section 1.6),

$$P_n(x) = \Big(\cdots((a_n x + a_{n-1})x + a_{n-2})x + \cdots + a_1\Big)x + a_0 \tag{8.20}$$

The laboratory participants can verify that (8.20) requires merely n multiplications and n additions, or exactly $2n$ computing units. It follows that a slight rearrangement of terms has saved us about 33 percent of the computational effort.

At this point the instructor can raise the following questions: Is it possible to perform preliminary (one-time) computations that cast $P_n(x)$ into an even more efficient form? Is there a most efficient form? Such questions relate to the theory of *computational complexity*. Here, we offer only a glimpse into the achievements of this theory. Let us take a fourth-order polynomial in Horner's form

$$P_4(x) = \Big(((a_4 x + a_3)x + a_2)x + a_1\Big)x + a_0 \tag{8.21}$$

which consumes eight computing units per evaluation. The theory of computational complexity would have us rewrite (8.19) in the form

$$P_4(x) = a_4\big[x(x+\alpha)+\beta\big]\big[x(x+\alpha)+x+\gamma\big]+\delta \tag{8.22}$$

which looks somewhat strange at first. The laboratory participants may convince themselves of the fact that α, β, γ, and δ are determined uniquely by a_0, a_1, a_2, a_3, and a_4, and conversely (see Exercise 8.10). Again, the computations of α, β, γ, and δ, however time-consuming, are carried out only once. In (8.22), it is necessary to evaluate the term $x(x+\alpha)$ only once and thus (8.22) requires a total of three multiplications and five additions (hardly a gain in efficiency). However, for higher values of n, the corresponding result of the theory of computational complexity is that the evaluation of $P_n(x)$ via a generalization of (8.22) requires $(n+1)$ additions and approximately $(n/2+1)$ multiplications. Thus, only $(3/2)n$ computing units (approximately) are needed compared with $2n$ in Horner's form. This result is due to Donald Knuth, who discovered it in 1962.

Another remark is in order. In the construction of library functions, we are not limited to polynomials. We could just as well employ rational functions (quotients of polynomials) and proceed in a way analogous to what we have done. These rational approximations can then "compete" for efficiency (for a given tolerable error) with corresponding polynomial approximations. We look at this subject next.

8.5 Potential of rational approximations

Now that we have constructed polynomial approximations to functions to be installed permanently in the computer's library, we naturally ask whether we can make use of divisions, as well as multiplications and additions, for the construction of library functions. In other words, why not use rational approximations, that is, quotients of polynomials, for that purpose. We shall do just that, by means of an example that demonstrates the computational potential of rational approximations. Specifically, we shall construct rational approximations of $f(x) = \ln x$, and as we saw earlier, it is sufficient to consider only the interval $[1,2]$. Because our sole intention at this point is to demonstrate the workings of rational approximations, we ignore the issue of relative-error control discussed earlier, and apply our methods to $f(x) = \ln x$ rather than to $g(x) = \ln x/(x-1)$.

To illustrate our ideas, we limit our discussion to three-parameter approximations based on three interpolation points. For the purpose of comparison, we shall also incorporate a parabolic (i.e., a second-degree polynomial) approximation. The rational approximations with three parameters can be represented in one of the following forms:

$$\text{parabola} \qquad (px+q)x+r \tag{8.23}$$

$$\text{hyperbola} \qquad \alpha + \frac{\beta}{x+\gamma} \tag{8.24}$$

$$\text{“inverse” parabola} \qquad \frac{a}{(x+b)x+c} \tag{8.25}$$

In each form the sum of the degrees of the numerator and the denominator is 2. All three are represented in a computationally economic form. The evaluation of the first involves two additions and two multiplications; the second, two additions and one division; and the third, two additions, one multiplication, and one division. In all those approximations, three interpolation points and four *critical points*, points of maximal deviation from $f(x)$, are expected. The equal ripple approximation in each of the three forms cannot be found analytically, so we adopt the interpolative approach. In this approach, we will construct approximations that coincide with the relevant function at three prescribed points. These points are chosen from a table of abscissas at which the values of the relevant function are accurately known. To do this, let us examine the function $y = \ln x$ in the interval $[1, 2]$. For the preparation of this table, we calculate once the constant $t = 2^{1/32}$ by five successive square-root extractions [see also equation (7.46) and the discussion that follows]. In addition, we compute the constant $z = (\ln 2)/32$ to a desirable accuracy. Now, the x-values in the table will be $1, t, t^2, \ldots, t^{32}$ and the corresponding known values of $y = \ln x$ are $0, z, 2z, \ldots, 32z$. From this table we select our three suitable interpolation points. At first, it would seem natural to choose initial interpolation abscissas that are close to the equidistant points 1.25, 1.50, and 1.75. However, from our experience in the previous chapter we know that it is preferable to choose the first and last abscissas closer to the interval endpoints. Hence, we recommend choosing initial interpolation points (from the table) whose x-values are nearly equal to 1.1, 1.5, and 1.9. From Table 8.1 it is clear that the initial values of x should be

$$x_0 = 1.090508, \qquad x_1 = 1.509164, \qquad x_2 = 1.915206 \tag{8.26}$$

We start our three-parameter approximations with the parabola (8.23),

Table 8.1. *Thirty-three values of* $\ln x$ *in* $[1, 2]$

j	$x = t^j$	$y = jz$
0	1.000000	0.000000
1	1.021897	0.021661
2	1.044274	0.043322
3	1.067141	0.064983
4	1.090508	0.086643
5	1.114387	0.108304
6	1.138789	0.129965
7	1.163725	0.151626
8	1.189207	0.173287
9	1.215248	0.194948
10	1.241858	0.216608
11	1.269052	0.238269
12	1.296840	0.259930
13	1.325237	0.281591
14	1.354256	0.303252
15	1.383911	0.324913
16	1.414214	0.346574
17	1.445182	0.368234
18	1.476827	0.389895
19	1.509164	0.411556
20	1.542212	0.433217
21	1.575982	0.454878
22	1.610492	0.476539
23	1.645757	0.498199
24	1.681794	0.519860
25	1.718621	0.541521
26	1.756254	0.563182
27	1.794711	0.584843
28	1.834010	0.606504
29	1.874170	0.628165
30	1.915206	0.649825
31	1.957146	0.671486
32	2.000000	0.693147

$px^2 + qx + r$. The collocation of this parabola with $\ln x$ at (x_0, y_0), (x_1, y_1), and (x_2, y_2), requires (see Exercise 8.11)

$$\begin{aligned} p &= \frac{\frac{y_2 - y_1}{x_2 - x_1} - \frac{y_1 - y_0}{x_1 - x_0}}{x_2 - x_0} \\ q &= \frac{y_1 - y_0}{x_1 - x_0} - p(x_0 + x_1) \\ r &= y_0 - \frac{y_1 - y_0}{x_1 - x_0} x_0 + p x_0 x_1 \end{aligned} \tag{8.27}$$

As pointed out, we started with x_0, x_1, and x_2 given in (8.26); these are the points 4, 19, and 30 in Table 8.1. Next, we ran a program tabulating the differences between the $\ln x$ of the computer's library and our parabolic approximation. These differences represent the error function for values of x increasing from 1 to 2, with a desired increment (e.g., 0.005). This table was accompanied by a corresponding graph (see Exercise 8.12).

After inspecting the resulting table, we changed and rechanged the interpolation points (using Table 8.1) so as to get smaller error ripples. Thus, the resulting approximation gradually approached the state of having the equal ripple property with *four critical points*: two in the interior and two being the endpoints of the relevant interval (at $x = 1$ and $x = 2$ in our case). The best interpolation points (in the sense of near-equal ripples) in Table 8.1 were thus found to be points no. 2, 17, and 30 because their error ripples turned out closest to being equal. With these best interpolation points, the values of the error function

$$E_P(x) = \ln x - (px^2 + qx + r) \tag{8.28}$$

at the *four critical points* on the resulting graph, were found (see Exercise 8.13) to be

$$\begin{aligned} E_P(1.00) &= -0.0027 \\ E_P(1.20) &= +0.0035 \\ E_P(1.72) &= -0.0036 \\ E_P(2.00) &= +0.0038 \end{aligned} \tag{8.29}$$

Inspecting (8.29), we decided to further improve the uniformity of the error ripples by using interpolation points from a denser table, generated with $t = 2^{1/128}$ and $z = (\ln 2)/128$ as in (7.46). Because this table is quite long, we show only its relevant parts in Table 8.2.

Using this new table, we continued to improve the approximation. The best interpolation points in Table 8.2, found in this manner, are points no. 9, 69, and 121. The resulting error function was found (see Exercise 8.14) to have the four extremal values

$$\begin{aligned} E_P(1.00) &= -0.0030 \\ E_P(1.22) &= +0.0035 \\ E_P(1.72) &= -0.0036 \\ E_P(2.00) &= +0.0033 \end{aligned} \tag{8.30}$$

Table 8.2. *Denser values of* $\ln x$ *in* $[1, 2]$

j	$x = t^j$	$y = jz$
0	1.000000	0.000000
⋮	⋮	⋮
6	1.033025	0.032491
7	1.038634	0.037906
8	1.044273	0.043322
9	1.049944	0.048737
10	1.055645	0.054152
⋮	⋮	⋮
63	1.406572	0.341158
64	1.414209	0.346574
65	1.421888	0.351989
66	1.429609	0.357404
67	1.437371	0.362819
68	1.445176	0.368234
69	1.453023	0.373650
70	1.460913	0.379065
71	1.468845	0.384480
72	1.476821	0.389895
73	1.484840	0.395310
⋮	⋮	⋮
118	1.894565	0.638995
119	1.904853	0.644410
120	1.915196	0.649825
121	1.925595	0.655241
122	1.936050	0.660656
⋮	⋮	⋮
128	2.000000	0.693147

Obviously, these results can be further improved by using even denser interpolation points. However, (8.30) suffices for our purposes, as will be seen in Fig. 8.1.

The deviations in (8.30) give a clear indication of the quality of parabolic approximation to $\ln x$ in $[1, 2]$. Finally, the approximating parabola with error ripples given by (8.30) takes the form

$$P(x) = -0.240035x^2 + 1.406915x - 1.163814 \tag{8.31}$$

We now turn our attention to the construction of a rational approximation of the form (8.24), $\alpha + \beta/(x + \gamma)$. By requiring this hyperbola to coincide with $y = \ln x$ at (x_0, y_0), (x_1, y_1), and (x_2, y_2), we get the formulas for the coefficients

$$\begin{aligned}
\gamma &= \frac{(x_2 - x_1)(x_1 y_1 - x_0 y_0) - (x_1 - x_0)(x_2 y_2 - x_1 y_1)}{(y_2 - y_1)(x_1 - x_0) - (y_1 - y_0)(x_2 - x_1)} \\
\alpha &= \frac{(y_2 - y_1)(x_1 y_1 - x_0 y_0) - (y_1 - y_0)(x_2 y_2 - x_1 y_1)}{(y_2 - y_1)(x_1 - x_0) - (y_1 - y_0)(x_2 - x_1)} \\
\beta &= (y_0 - \alpha)(x_0 + \gamma)
\end{aligned} \tag{8.32}$$

As our initial interpolation points, we chose points no. 9, 69, and 121 from Table 8.2, which gave us the best parabolic approximation. After improving this hyperbolic approximation repeatedly, using Table 8.2, we found the best points in the table to be points no. 9, 64, and 119. The hyperbolic approximation with these points was found to take the form

$$R(x) = 2.361334 - \frac{5.698596}{x + 1.414210} \tag{8.33}$$

with extremal error values (see Exercise 8.15) of

$$\begin{aligned}
E_R(1.00) &= -0.0009 \\
E_R(1.19) &= +0.0008 \\
E_R(1.68) &= -0.0008 \\
E_R(2.00) &= +0.0009
\end{aligned} \tag{8.34}$$

Thus, the maximal error of our rational–hyperbolic approximation, with almost equal error ripples, is about $\frac{1}{4}$ of the corresponding maximal error of the polynomial–parabolic approximation. This result demonstrates the potential of rational approximations, which will be discussed further later.

For completeness, we repeated the entire process with the approximation of the form (8.15), $a/(x^2+bx+c)$. In this case, even after improving the choice of interpolation points, the size of the error ripples was about 30 times that of the parabolic approximation. Thus, we have decided to exclude the error function corresponding to this third approximation from Fig. 8.1, in which we display

$$\begin{aligned}
E_P(x) &= \ln x - P(x) \\
E_R(x) &= \ln x - R(x)
\end{aligned} \tag{8.35}$$

where $P(x)$ and $R(x)$ are given in (8.31) and (8.33), respectively.

The superiority of rational approximations over polynomial ones, demonstrated here for $\ln x$, is not just a lucky break. Corroborating evidence

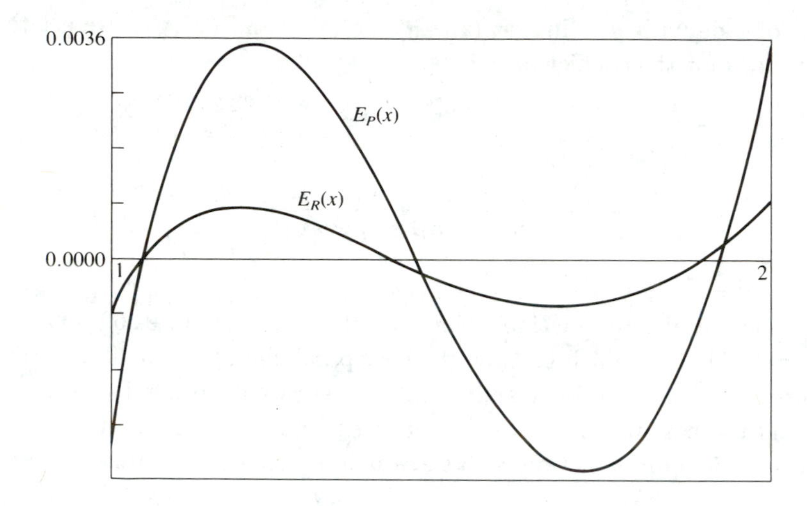

Fig. 8.1

for the computational potential of rational approximations will be given in the next section, where we shall also see that the "superiority-factor" increases with the number of parameters (i.e., interpolation points) employed in the approximations.

8.6 Polynomial versus rational approximation

To make a fair comparison of rational with polynomial approximations for a function $f(x)$ in an interval $[a, b]$, so as to assess their relative qualities, both approximations should possess the same number of parameters, that is, the same number of interpolation points. Indeed, if the rational approximation is $R_{m,n}(x)$, whose numerator is a polynomial of degree m and whose denominator is a polynomial of degree n, then the corresponding polynomial approximation should be $P_{m+n}(x)$, of degree $(m+n)$. Moreover, the computational effort (i.e., the computing time) of evaluating either $R_{m,n}(x)$ or $P_{m+n}(x)$ at a given x is about the same (see Exercise 8.16).

It should be pointed out that a simple formula for the construction of $R_{m,n}(x)$ for given $(m+n+1)$ interpolation points – analogous to the Lagrangian formula (7.4)–(7.5) discussed in the previous chapter – *is not available*. Thus, Section 8.5, where we employed three interpolation points, we actually used "brute force" to determine the coefficients

α, β, and γ of the rational approximation (8.24) and eventually obtained (8.32). This procedure becomes ever more cumbersome as the number of interpolation points is increased. Nevertheless, we did carry it out for five interpolation points and for various functions, and constructed the corresponding $P_4(x)$ and $R_{2,2}(x)$ to gain further insight into the competition between rational and polynomial approximations. Moreover, we juggled the interpolation points for each approximation so as to obtain nearly equal error ripples.

The first function we approximated, with nearly equal error ripples, was $f(x) = \sqrt{x}$ in $[1,4]$, for which we found

$$\operatorname*{Max}_{1\le x\le 4} \left|\sqrt{x} - P_4(x)\right| = 0.000370 \tag{8.36}$$

and

$$\operatorname*{Max}_{1\le x\le 4} \left|\sqrt{x} - R_{2,2}(x)\right| = 0.000028 \tag{8.37}$$

that is, the rational approximation is superior by a factor of about 13. We note in passing that we also tried the rational approximations $R_{3,1}(x)$, $R_{1,3}(x)$, and $R_{0,4}(x)$, but they turned out to be inferior to $R_{2,2}(x)$. Figure 8.2 displays the pointwise comparison between $E_P(x) = \sqrt{x} - P_4(x)$ and $E_R(x) = \sqrt{x} - R_{2,2}(x)$, in which we also see the nearly equal error ripples and the (different) interpolation points of both approximations.

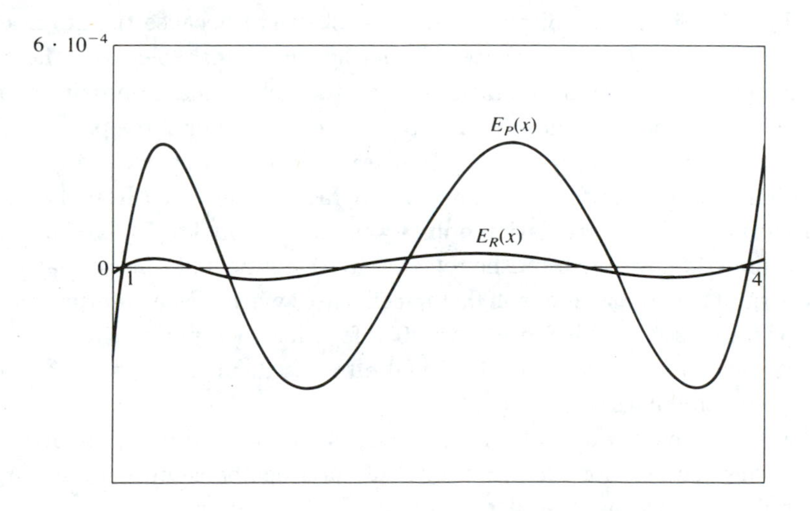

Fig. 8.2

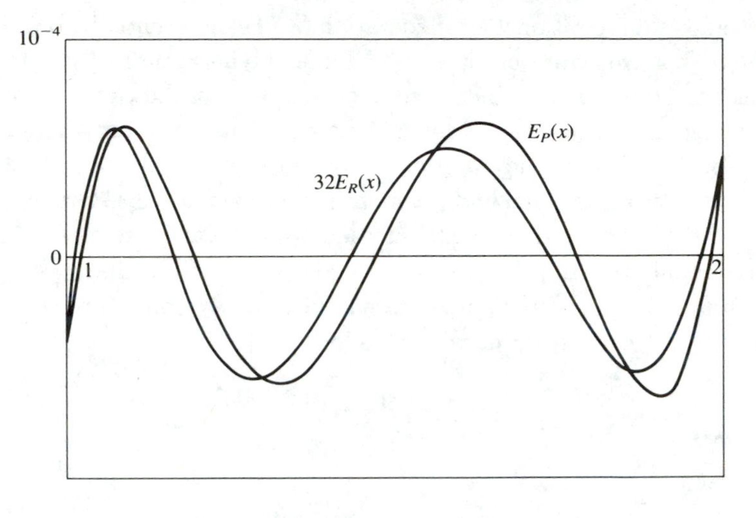

Fig. 8.3

As our second example, we constructed $P_4(x)$ and $R_{2,2}(x)$ for $f(x) = \ln x$ in $[1, 2]$ and found

$$\operatorname*{Max}_{1\le x\le 2} \left| \ln x - P_4(x) \right| = 0.000064 \tag{8.38}$$

$$\operatorname*{Max}_{1\le x\le 2} \left| \ln x - R_{2,2}(x) \right| = 0.000002 \tag{8.39}$$

For this case the rational approximation is superior by a factor of 32. Indeed, in Fig. 8.3 we display $E_P(x)$ vs. $32E_R(x)$ because the graph of $E_R(x)$ itself, in the scale chosen, is hardly distinguishable from the x-axis. The superiority of the rational over the polynomial approximation in this case is striking indeed. Moreover, if we recall our three-parameter approximations for $f(x) = \ln x$, whose error ripples are shown in Fig. 8.1, we immediately see that the "superiority-factor" increased from 4 to 32 as the number of interpolation points was raised from 3 to 5. For the sake of completion, we cite from the relevant mathematical literature that the superiority-factor is about 400, for this case, when seven interpolation points are used, and is more than 4000 for nine such points.

We have treated several additional often-used functions, and the results are summarized in Table 8.3.

The table clearly indicates that when polynomial and rational approximations compete for the "privilege" of entering the computer's library of built-in functions, the latter usually puts the former in the shade. An exceptional case, however, at least for five interpolation points, is $\sin x$,

Table 8.3. *Comparison of polynomial and rational approximations for five interpolation points*

$f(x)$	$[a,b]$	Max$\lvert E_P(x)\rvert$	Max$\lvert E_R(x)\rvert$	Ratio $\approx$
$\sqrt{x}$	$[1,4]$	0.000370	0.000028	13
$\ln x$	$[1,2]$	0.000064	0.000002	32
$\arctan x$	$[0,10]$	0.040720	0.001508	27
$\tan(\pi x/2)$	$[0,0.5]$	0.000288	0.000017	17
$\sin(\pi x/2)$	$[0,1]$	0.000110	0.000202	$\frac{5}{9}$
e^x	$[-1,1]$	0.000547	0.000087	6

for which the polynomial approximation is the winner. In general, many questions concerning polynomial versus rational approximation are still unanswered; thus, in the absence of appropriate theorems we must rely on computational experimentation.

A final remark is in order. So far we have endeavored to reduce the interval over which we approximate the given function as much as possible, in order to enhance computational efficiency. When the final interval $[a,b]$ is still too large, in the sense that the approximation must be of an unreasonably high degree in order to attain library accuracy, we can artificially split that interval into a number of subintervals. In each subinterval we then employ a different approximation of a reduced degree. This being the case, one might wonder about splitting even a small interval of approximation. For example, the interval $[1,2]$ over which we have approximated $\ln x$ could be split, say, into the subintervals $[1.0,1.5]$ and $[1.5,2.0]$. In each of these subintervals we could construct a different approximation of a lower degree (and thus more efficient) than the original one, carrying the same accuracy. This is indeed possible and should be tried computationally. In this connection see Exercises 8.17 and 8.18. This idea – known as *segmentation* – should not, however, be carried too far. By excessive segmentation we can end up, say, with a built-in function consisting of several hundreds of parabolas, which will burden the computer's permanent memory intolerably. The situation can be likened to a carpenter carrying four screwdrivers of various sizes, which makes practical sense, but it would certainly be impractical to carry several hundred screwdrivers so as to fit precisely every conceivable screw. For the construction of computer library functions, a middle road must be found, balancing the desire to save computing time (using polynomials of moderate degree) against the need to conserve permanent memory space (using limited segmentation). Similar compromises must be made in many practical applications of numerical mathematics, as has been demonstrated repeatedly throughout these laboratory assignments.

In concluding this book, we take a final overview of the material it covers through its unique approach. This book introduces computational microcomputer laboratories as a vehicle for teaching algorithmic aspects of mathematics. This is achieved through a sequence of laboratory assignments, presupposing no previous knowledge of calculus or linear algebra, where the chalk-and-talk lecturer turns into a laboratory instructor. We believe the material in this book is part and parcel of the mathematical foundations that should be acquired by a college student facing the twenty-first century.

Exercises

8.1 Write a program that for any input t, $-\infty < t < \infty$, outputs x, $0 \leq x \leq \pi/2$, and $s = 1$ or $s = -1$, so that $\cos t = s \cdot \cos x$. Observe that this constitutes the preparatory part of a built-in cosine function routine.

8.2 Write and run a program that generates the collection T of the 128 values $\cos t$, $\cos 2t, \ldots, \cos 128t$, where $t = (1/128)(\pi/2)$, as described by means of equations (8.2) and (8.4). Make sure to use double precision in the calculations, but print the results with single precision. For the purpose of comparison, repeat the calculations using single precision throughout, and observe the numerical contamination caused by round-off errors.

8.3 Study the accompanying figure and compare the areas of the triangle ABC, the circular sector ABE, and the triangle ADE, where the angle α is measured in radians.

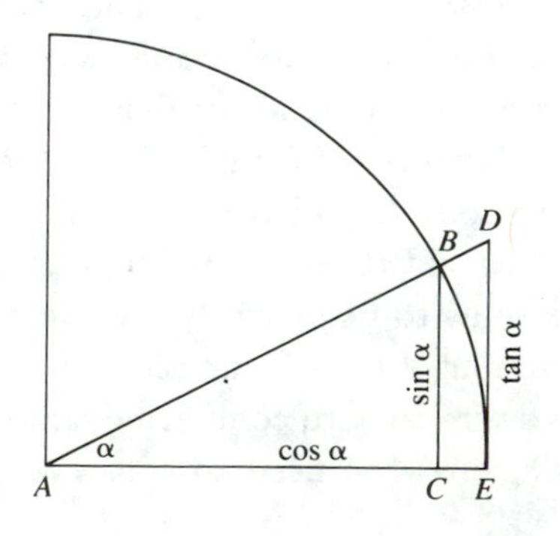

Conclude that

$$\frac{1}{2}\sin\alpha\cos\alpha \le \frac{1}{2}\alpha \le \frac{1}{2}\tan\alpha$$

and hence

$$\cos\alpha \le \frac{\alpha}{\sin\alpha} \le \frac{1}{\cos\alpha}$$

Now let the acute angle α decrease toward zero, and observe that because $\cos\alpha$ increases toward unity while $1/\cos\alpha$ decreases toward unity, the quantity $\alpha/\sin\alpha$ (being sandwiched between) approaches unity, and so does $(\sin\alpha)/\alpha$.

8.4 Multiply the first inequality in Exercise 8.3 by 4, set $x = 2\alpha$, and thus conclude that $\sin x \le x$ for $x \ge 0$. Next draw the graphs of $y = x$ and $y = \sin x$, in the interval $[0, \pi/2]$, on one set of axes. Consider the quotient $(\sin x)/x$, observe the increasing gap between the values of the denominator and the numerator, and deduce that the quotient decreases from unity at $x = 0$ to $2/\pi \approx 0.6366$ at $x = \pi/2$ (which means, of course, that $x/\sin x$ increases from 1 to $\pi/2 \approx 1.5708$).

8.5 Consider the quotient $\ln x/(x-1)$ for values of x decreasing toward unity. Set $u = 1/(x-1)$ and observe that as x decreases toward unity, u increases indefinitely. Moreover, we have

$$\frac{\ln x}{x-1} = u\ln\left(1+\frac{1}{u}\right) = \ln\left(1+\frac{1}{u}\right)^u$$

Now recall the results presented in Section 5.8 leading to the number e, and conclude that as x decreases toward unity, the quotient $\ln x/(x-1)$ increases toward unity. As a consequence, deduce that $\ln x \le x - 1$.

8.6 Bearing in mind that $\ln x \le x - 1$, draw the graphs of $y = x - 1$ and $y = \ln x$, in the interval $[1, 2]$, on one set of axes. Consider the quotient $\ln x/(x-1)$, observe the increasing gap between the values of the denominator and the numerator, and deduce that the quotient decreases from unity at $x = 1$ to $\ln 2 \approx 0.6931$ at $x = 2$ (which means, of course, that $(x-1)/\ln x$ increases from 1 to $1/\ln 2 \approx 1.4427$).

8.7 Suppose that the function $f(x)$ that we want to approximate possesses a root of multiplicity p_1 at $x = r_1$ and a root of multiplicity p_2 at $x = r_2$, both in the relevant interval. Which function $g(x)$ should be approximated in order to control the relative error in the approximation of $f(x)$ throughout that interval? What form will (8.16) take in this case?

8.8 Suppose we want to approximate $f(x)$ in $[a, b]$ and that $1/f(x)$ has a simple root at $x = s$ in this interval (such a point is called a *simple pole*). Show that if we approximate the function $g(x) = (x - s)f(x)$ by a polynomial $P_n(x)$ and subsequently take $P_n(x)/(x-s)$ as the approximation to $f(x)$, then we are in control of the relevant relative error. Generalize this procedure for a pole of order p.

8.9 The function $\tan x$ has a simple root at $x = 0$ and a simple pole at $x = \pi/2$ (verify this fact). Which function $g(x)$ should be approximated by $P_n(x)$ so as to control the relevant relative error? What is the form of the approximation to $\tan x$?

8.10 The evaluation of $P_4(x) = a_4x^4 + a_3x^3 + a_2x^2 + a_1x + a_0$ for a given value of x requires seven multiplications and four additions, compared with four multiplications and four additions in Horner's form $P_4(x) = (((a_4x + a_3)x + a_2)x + a_1)x + a_0$. Prove that the polynomial $P_4(x)$ can be cast into the form $P_4(x) = a_4[x(x + \alpha) + \beta][x(x + \alpha) + x + \gamma] + \delta$, and express α, β, γ, and δ in terms of a_0, a_1, a_2, a_3, and a_4.

8.11 Show that if the parabola $(px + q)x + r$ is to pass through the points (x_0, y_0), (x_1, y_1), and (x_2, y_2), then p, q, and r are given by (8.27). Compare the resulting parabola with the parabola given in the Lagrangian form (7.3), show that they are equivalent, and explain why.

8.12 Write and run a program that prints out and graphs the differences between the values of $\ln x$ given by the computer's library function and those of the parabolic approximation based on (8.27) and (8.26). Have x increase from 1 to 2, using increments of 0.005.

8.13 Repeat Exercise 8.12, using interpolation points no. 2, 17, and 30 in Table 8.1. Compare the resulting graph of the difference $\ln x - (px^2 + qx + r)$ with the one obtained in Exercise 8.12, and observe that the new error ripples are closer to being equal than the previous ones.

8.14 Repeat Exercise 8.12 once again, this time using the interpolation points no. 9, 69, and 121 in Table 8.2. Observe the further improvement of the error ripples (relative to those in Exercise 8.13).

8.15 Write and run a program that prints out and graphs the differences between the values of $\ln x$ given by the computer's library function and those of the hyperbolic approximation based on

(8.24) with the points no. 9, 64, and 119 in Table 8.2. Let x increase from 1 to 2, with increments of 0.005, as in the previous exercises. Observe the almost equal error ripples, whose size is about a quarter of that obtained in Exercise 8.14 for the parabolic approximation.

8.16 How many parameters does a polynomial $P_{m+n}(x)$ possess, and through how many arbitrary interpolation points can this polynomial be passed? Convince yourself that these two numbers are the same and denote their common value by k. Answer the same questions for a rational function $R_{m,n}(x)$ and convince yourself that the corresponding two numbers are again equal. Is their common value equal to k or $(k+1)$? Explain.

8.17 Write and run a program that outputs Table 8.2 in its entirety, displaying all 129 possible interpolation points for $\ln x$ in $[1,2]$. In your calculations, use $\ln 2 = 0.6931472$ and $t = 2^{1/128} = 1.0054299$.

8.18 The approximation of $\ln x$ in the interval $[1,2]$ by a fourth-degree polynomial $P_4(x)$, with nearly equal error ripples, yielded a maximal error of 0.000064 as given in (8.38). Split the interval into several subintervals of equal length, so that comparable accuracy can be attained using just third-degree polynomials (one for each subinterval). Start with two subintervals, and perform the necessary computations using points from the table you generated in Exercise 8.17. Are two subintervals sufficient? If not, continue the segmentation.

8.19 Repeat Exercise 8.18, using parabolic (rather than cubic) approximation in each subinterval.

Solutions to selected exercises

Solutions to exercises calling for extensive computer printouts (such as tables and graphs) are not given here. The algorithms given in some solutions may of course be different if the specific programming language employed allows the use of if-then-else, while-loops, recursion, and so forth.

Chapter 1

1.1 Algorithm for $n!$

1. Input the value of n (a positive integer or zero).
2. Set $k = 0$ and $f = 1$.
3. If $k = n$, go to Step 7.
4. Replace the value of k by $(k + 1)$.
5. Replace the value of f by $(k \cdot f)$.
6. Return to Step 3.
7. Print the values of n and f.
8. End.

1.2 Algorithm for $(1 + 1/1! + 1/2! + \cdots + 1/n!)$

1. Input the value of n (positive integer or zero).
2. Set $k = 0,\ t = 1,\ s = 1$.
3. If $k = n$, go to Step 8.
4. Replace the value of k by $(k + 1)$.
5. Replace the value of t by (t/k).
6. Replace the value of s by $(s + t)$.
7. Return to Step 3.
8. Print the values of n and s.
9. End.

1.3 Using the formula $x = (-b \pm \sqrt{b^2 - 4ac})/(2a)$ for the solution of $ax^2 + bx + c = 0$, calculate (corresponding to $b = -2,\ c = -1$) the values $x_1 = \left(1 + \sqrt{1+a}\right)/a$ and $x_2 = \left(1 - \sqrt{1+a}\right)/a$, for $a = 0.1, 0.01, 0.001, \ldots, 0.0000001$. In each case, substitute x_1 and x_2 thus obtained in the quadratic equation, and see how much the values you obtain deviate from zero. Note that your numerical results depend on the intrinsic accuracy of the particular computing device you are using. Note also that the relative errors in the values of x_1 are small, but for x_2 they become progressively larger with decreasing a.

1.4 The results obtained, using a PC with a seven-digit intrinsic accuracy, are as follows:

q	$x_1/10^q$	x_2
2	1.4	2.499998
3	1.4	2.499998
4	1.4	2.499819
5	1.4	2.497435
6	1.4	2.503395
7	1.4	2.384186
8	1.4	5.960465
9	1.4	0
10	1.4	0
11	1.4	0
12	1.4	0

The errors in x_2 stem from the subtraction of two almost equal numbers in its numerator. By contrast, the computation of $x_2 = c/(ax_1)$ does not suffer from this numerical ailment.

1.5 The numbers of significant digits are 6, 5, 4, 5, 5, 4, 8, and 3.

1.6 $1.70305 \cdot 10^3$, $-6.3004 \cdot 10^1$, $7.003 \cdot 10^{-3}$, $-1.0101 \cdot 10^2$, $1.7300 \cdot 10^1$, $1.433 \cdot 10^{-6}$, $6.2013041 \cdot 10^6$, and $3.48 \cdot 10^0$

1.7 Suppose the number x is given in scientific notation in the form $x = a \cdot 10^q = b \cdot 10^p$, where p and q are integers, $1 \le a < 10$, and $1 \le b < 10$. Without loss of generality we may assume $p \ge q$. Multiplying by 10^{-q} we obtain $a = b \cdot 10^{p-q}$. Now if $p = q$, we immediately have $a = b$ as well. If $p > q$, then $p - q \ge 1$; and so $1 \le b < 10$ implies $a > 10$, contradicting $1 \le a < 10$. Note that the same proof holds for x expressed in any base, not just 10.

1.8 10, 0.1, 10000, 111.01, 10101, 0.001, and $0.0001100110\overline{011}$

1.9 5.375, 1988, 3.75, 0.125, 7, 9, and 0.2

1.10 Multiplying out both expressions for $P_4(x)$, and comparing coefficients of like powers, one obtains four equations for the four unknowns α, β, γ, and δ. The solution of these equations leads to the unique expressions

$$\alpha = \frac{a_3 - a_4}{2a_4}$$

$$\beta = \frac{(a_4 - a_3)t + 8a_4^2 a_1}{8a_4^3}$$

$$\gamma = \frac{(a_4 + a_3)t - 8a_4^2 a_1}{8a_4^3}$$

$$\delta = a_0 - \frac{(a_4^2 - a_3^2)t^2 + 16a_1 a_3 a_4^2 t - 64a_1^2 a_4^4}{64a_4^5}$$

where $t = a_4^2 - a_3^2 + 4a_2a_4$ and, of course, $a_4 \neq 0$.

Chapter 2

2.1 Given $d = m \cdot 2^i$, where $1 \leq m < 2$ and i is an appropriate integer. There are three cases: (a) The integer i is divisible by 3. Denoting $j = i/3$, we have $d = s \cdot 8^j$, where $s = m$. (b) When divided by 3, the integer i leaves the remainder one; that is, $i = 3j + 1$ for some integer j. Thus, $j = (i-1)/3$, and we have $d = m \cdot 2^i = (2m) \cdot 2^{i-1} = (2m) \cdot 8^{(i-1)/3}$; that is, $d = s \cdot 8^j$ where $s = 2m$. (c) When divided by 3, the integer i leaves the remainder two; that is, $i = 3j + 2$ for some integer j. Thus, $j = (i-2)/3$ and we have $d = m \cdot 2^i = (4m) \cdot 2^{i-2} = (4m) \cdot 8^{(i-2)/3}$; that is, $d = s \cdot 8^j$ where $s = 4m$. In all cases, because $1 \leq m < 2$, we have $1 \leq s < 8$, so $1 \leq \sqrt[3]{s} < 2$. The uniqueness of the representation $d = s \cdot 8^j, 1 \leq s < 8$, follows from Exercise 1.7.

2.2 Repeat the arguments of Exercise 2.1, this time considering k cases according to whether i leaves the remainder 0, 1, 2, ... $(k-2)$, or $(k-1)$ when divided by k.

2.4 Rewrite the equation $2^x + x - 5 = 0$ as $2^x = 5 - x$, and draw the graphs of $u = 2^x$ and $v = 5 - x$ in the same coordinate system. From this deduce that the two curves intersect exactly once, and because $u(1) = 2 < 4 = v(1)$ and $u(2) = 4 > 3 = v(2)$, it follows that the point of intersection is located between $x = 1$

and $x = 2$. The bisection algorithm resembles the one in Section 2.3 with $y = 2^x + x - 5$.

2.5 Consider $x^7 = 200 - 2x$, and then follow the arguments of Exercise 2.4.

2.6 Trisection algorithm for $\sqrt{s}$, $1 \le s < 4$.

1. Input the value of q (desired number of correct figures).
2. Set $a = 1$, $b = 2$, and $E = \frac{1}{2} \cdot 10^{-q}$.
3. Input the value of s ($1 \le s < 4$, after range reduction).
4. Compute $x = (2a + b)/3$ and $y = x^2 - s$.
5. If $y = 0$, go to Step 11.
6. If $y > 0$, then $b = x$ and go to Step 10.
7. Compute $x = (a + 2b)/3$ and $y = x^2 - s$.
8. If $y = 0$, go to Step 11.
9. If $y > 0$, then $b = x$ and $a = (a + x)/2$, else $a = x$.
10. If $(b - a) > E$, return to Step 4.
11. Output the values of x and y.
12. End.

For a large number of computations (for different values of s), there is a chance of $\frac{1}{3}$ that the root is in the leftmost third of the interval $[a, b]$, and hence only one evaluation of $y = x^2 - s$ is needed per iteration. Otherwise, two evaluations are needed. It follows that the average number of evaluations per iteration equals $\frac{1}{3} \cdot 1 + \frac{2}{3} \cdot 2 = \frac{5}{3}$.

2.7 Because the bisection and the trisection yielded the same accuracy, we have $(b - a)/2^n \approx (b - a)/3^m$, and thus $3^m \approx 2^n$, or $m/n \approx \log 2/\log 3$. Denote the computational work per evaluation by w, so that

$$\frac{\text{total work in trisection}}{\text{total work in bisection}} = \frac{m \cdot \dfrac{5}{3} \cdot w}{n \cdot 1 \cdot w} \approx \frac{5 \log 2}{3 \log 3} = 1.0515$$

Accordingly, bisection enjoys a 5.15 percent edge over trisection.

2.8 Since $1/2^{10} = 1/1024 < 10^{-3}$, we have for $n \ge \frac{10}{3}q + 1$ that

$$\begin{aligned} \frac{1}{2^n} = \left(\frac{1}{2}\right)\left(\frac{1}{2^{n-1}}\right) &= \left(\frac{1}{2}\right)\left(\frac{1}{2^{10}}\right)^{(n-1)/10} \\ &< \frac{1}{2} \cdot 10^{-(3/10)(n-1)} \\ &\le \frac{1}{2} \cdot 10^{-(3/10)(10/3)q} = \frac{1}{2} \cdot 10^{-q} \end{aligned}$$

2.9 Since $\log_{10} 2 > 0.3$ and we are given that $m \geq (1/0.3)\log_{10} n$, we have $m > \log_{10} n / \log_{10} 2 = \log_2 n$, hence $2^m > n$, and thus $(\frac{1}{2})^{2^m} < (\frac{1}{2})^n$.

2.10 From (2.15) we know that the second-order iterations satisfy the condition $|x_n - \sqrt{s}| \leq \frac{1}{2}|x_{n-1} - \sqrt{s}|^2$. Moreover, from (2.7) it follows that $|x_0 - \sqrt{s}| \leq \frac{1}{2}$. Therefore,

$$|x_1 - \sqrt{s}| \leq \frac{1}{2}|x_0 - \sqrt{s}|^2 \leq \frac{1}{2}\left(\frac{1}{2}\right)^2$$

$$|x_2 - \sqrt{s}| \leq \frac{1}{2}|x_1 - \sqrt{s}|^2 \leq \frac{1}{2}\left(\frac{1}{2}\right)^2\left(\frac{1}{2}\right)^4$$

$$|x_3 - \sqrt{s}| \leq \frac{1}{2}|x_2 - \sqrt{s}|^2 \leq \frac{1}{2}\left(\frac{1}{2}\right)^2\left(\frac{1}{2}\right)^4\left(\frac{1}{2}\right)^8$$

and so on, and in general [see (2.20) for example],

$$|x_n - \sqrt{s}| \leq \left(\frac{1}{2}\right)^{1+2+4+\cdots+2^n} = \left(\frac{1}{2}\right)^{2^{n+1}-1} = 2\left(\frac{1}{2}\right)^{2^{n+1}}$$

Moreover,

$$|x_n - \sqrt{s}| \leq 2\left[\left(\frac{1}{2}\right)^2\right]^{2^n} = 2\left(\frac{1}{4}\right)^{2^n} < \left(\frac{1}{2}\right)^{2^n}$$

showing that this error estimate is sharper than (2.9).

2.11 The secant line connecting $(1, 1)$ and (2.25, 1.50) is

$$y = \frac{1.50 - 1.00}{2.25 - 1.00}(s - 1) + 1 = \frac{2}{5}s + \frac{3}{5}$$

and similarly the secant line connecting (2.25, 1.50) and $(4, 2)$ is

$$y = \frac{2}{7}s + \frac{6}{7}$$

The first tangent line is $y = \frac{2}{5}s + \alpha$, with α determined by the condition that $\sqrt{s} = \frac{2}{5}s + \alpha$ has only one (double) solution. This yields $\alpha = \frac{5}{8}$ so that the tangent line is $y = \frac{2}{5}s + \frac{5}{8}$. Similarly, the second tangent line is $y = \frac{2}{7}s + \frac{7}{8}$. The first middle line is thus $y = \frac{2}{5}s + (\frac{3}{5} + \frac{5}{8})/2 = \frac{2}{5}s + \frac{49}{80}$, and similarly the second middle line is $y = \frac{2}{7}s + \frac{97}{112}$.

2.12 For $1 \le s \le 2.25$ and $x_0 = \frac{2}{5}s + \frac{49}{80}$, we have (see the figure you constructed in Exercise 2.11),

$$|x_0 - \sqrt{s}| \le \frac{1}{2}\left(\frac{5}{8} - \frac{3}{5}\right) = \frac{1}{80}$$

whereas for $2.25 \le s < 4$ and $x_0 = \frac{2}{7}s + \frac{97}{112}$, we get

$$|x_0 - \sqrt{s}| \le \frac{1}{2}\left(\frac{7}{8} - \frac{6}{7}\right) = \frac{1}{112} < \frac{1}{80}$$

Thus, $|x_0 - \sqrt{s}| \le \frac{1}{80}$ for all s in $[1, 4]$, and by (2.21) with $K = \frac{1}{2}$, we have $|x_n - \sqrt{s}| \le 2\left(\frac{1}{160}\right)^{2^n}$

2.13 From $|x_{n+1} - \sqrt{s}| \le |x_n - \sqrt{s}|^3$ and $|x_0 - \sqrt{s}| \le \lambda$, it follows that

$$|x_1 - \sqrt{s}| \le |x_0 - \sqrt{s}|^3 \le \lambda^3$$
$$|x_2 - \sqrt{s}| \le |x_1 - \sqrt{s}|^3 \le \lambda^9$$
$$|x_3 - \sqrt{s}| \le |x_2 - \sqrt{s}|^3 \le \lambda^{27}$$

and so on, and in general,

$$|x_n - \sqrt{s}| \le \lambda^{3^n}$$

2.14 Starting from

$$x_{n+1} - \sqrt{s} = \frac{(x_n - \sqrt{s})^4}{r_n}$$

we get

$$x_{n+1} = \frac{x_n^4 + 6x_n^2 s + s^2}{r_n} + \left(1 - \frac{4x_n^3 + 4x_n s}{r_n}\right)\sqrt{s}$$

and so we choose $r_n = 4x_n^3 + 4x_n s$, yielding

$$x_{n+1} = \frac{x_n^4 + 6sx_n^2 + s^2}{4x_n^3 + 4sx_n}$$

These iterations satisfy (by design)

$$x_{n+1} - \sqrt{s} = \frac{(x_n - \sqrt{s})^4}{4x_n^3 + 4sx_n}$$

Because $x_0 > 1$, it follows that $x_1 \ge \sqrt{s} \ge 1$, and indeed we have $x_n \ge \sqrt{s} \ge 1$ for $n = 1, 2, 3, \ldots$. Since now $x_n \ge 1$ for all n including $n = 0$, we have $4x_n^3 + 4sx_n \ge 8$, and thus

$$|x_{n+1} - \sqrt{s}| \le \frac{1}{8}|x_n - \sqrt{s}|^4$$

Now since $|x_0 - \sqrt{s}| \le \lambda$, we can write

$$|x_1 - \sqrt{s}| \le \frac{1}{8}|x_0 - \sqrt{s}|^4 \le \frac{1}{8}\lambda^4$$
$$|x_2 - \sqrt{s}| \le \frac{1}{8}|x_1 - \sqrt{s}|^4 \le \frac{1}{8}\left(\frac{1}{8}\right)^4 \lambda^{16}$$
$$|x_3 - \sqrt{s}| \le \frac{1}{8}|x_2 - \sqrt{s}|^4 \le \frac{1}{8}\left(\frac{1}{8}\right)^4 \left(\frac{1}{8}\right)^{16} \lambda^{64}$$

and so on, and in general,

$$\begin{aligned}|x_n - \sqrt{s}| &\le \left(\frac{1}{8}\right)^{1+4+16+\cdots+4^{n-1}} \lambda^{4^n} \\ &= \left(\frac{1}{8}\right)^{(4^n-1)/3} \lambda^{4^n} = \left(\frac{1}{2}\right)^{4^n-1} \lambda^{4^n} \\ &= 2\left(\frac{\lambda}{2}\right)^{4^n}\end{aligned}$$

2.16 Since $x_0 \ge \sqrt[3]{s}$, we see from (2.53) that $x_n \ge \sqrt[3]{s}$ for all n, including $n = 0$. Therefore,

$$\frac{2}{3x_n} + \frac{\sqrt[3]{s}}{3x_n^2} \le \frac{2}{3\sqrt[3]{s}} + \frac{\sqrt[3]{s}}{3\left(\sqrt[3]{s}\right)^2} = \frac{1}{\sqrt[3]{s}} \le 1$$

since $1 \le \sqrt[3]{s} < 2$, and thus $|x_{n+1} - \sqrt[3]{s}| \le |x_n - \sqrt[3]{s}|^2$. Now we use $0 \le x_0 - \sqrt[3]{s} \le \delta$ to obtain, recursively, $|x_1 - \sqrt[3]{s}| \le \delta^2$, $|x_2 - \sqrt[3]{s}| \le \delta^4$, ..., $|x_n - \sqrt[3]{s}| \le \delta^{2^n}$

2.17 In Exercise 2.16 we employed an upper initialization obtained by using the tangent line (2.66), which is parallel to the secant (2.58). Clearly, for $1 \le s \le 8$, the graph of $\sqrt[3]{s}$ is sandwiched between these two lines, and hence (2.67) holds. Therefore, if we choose the upper initialization $x_0 = \frac{1}{7}s + \frac{2}{3}\sqrt{\frac{7}{3}}$, the deviation $(x_0 - \sqrt[3]{s})$ is clearly nonnegative and satisfies

$$0 \le x_0 - \sqrt[3]{s} \le \frac{2}{3}\sqrt{\frac{7}{3}} - \frac{6}{7} = \frac{2\sqrt{21}}{9} - \frac{6}{7} = \delta$$

Note that δ is of course equal to twice the deviation in (2.70).

2.18 Carrying out the binomial expansions in (2.72) and regrouping, we obtain

$$x_{n+1} = \left(\frac{x_n^3 - s}{p_n} + \frac{x_n^4 - 4x_n s}{q_n} \right) + \left(1 - \frac{3x_n^2}{p_n} - \frac{4x_n^3 - s}{q_n} \right) \sqrt[3]{s} + \left(\frac{3x_n}{p_n} + \frac{6x_n^2}{q_n} \right) \left(\sqrt[3]{s} \right)^2$$

To get rid of $\sqrt[3]{s}$ and $\left(\sqrt[3]{s}\right)^2$, we require

$$\frac{3x_n^2}{p_n} + \frac{4x_n^3 - s}{q_n} = 1$$

$$\frac{3x_n}{p_n} + \frac{6x_n^2}{q_n} = 0$$

The unique solution of this system is $p_n = \left(2x_n^3 + s\right)/(2x_n)$, $q_n = -2x_n^3 - s$, which yields the iterative formula

$$x_{n+1} = \frac{1}{2} \left(x_n + \frac{3sx_n}{2x_n^3 + s} \right)$$

2.19 We substitute the expressions for p_n and q_n obtained in Exercise 2.18 into (2.72), and then factor out $\left(x_n - \sqrt[3]{s}\right)^3$. This yields

$$x_{n+1} - \sqrt[3]{s} = \left(x_n - \sqrt[3]{s}\right)^3 \left[\frac{2x_n}{2x_n^3 + s} - \frac{x_n - \sqrt[3]{s}}{2x_n^3 + s} \right]$$

or

$$x_{n+1} - \sqrt[3]{s} = \frac{x_n + \sqrt[3]{s}}{2x_n^3 + s} \left(x_n - \sqrt[3]{s}\right)^3$$

To complete the exercise, it remains to show that

$$0 < \frac{x_n + \sqrt[3]{s}}{2x_n^3 + s} \leq 1$$

The positivity follows readily from the iterative formula

$$x_{n+1} = \frac{1}{2} \left(x_n + \frac{3sx_n}{2x_n^3 + s} \right)$$

because always $x_0 \geq 1 > 0$. Thus, $x_0 \geq \sqrt[3]{s}$ (upper initialization) implies $x_n \geq \sqrt[3]{s}$ for all n. Similarly, for lower initialization, $x_n \le \sqrt[3]{s}$ for all n. For lower initialization, we rewrite the iterative formula in the form

$$x_{n+1} = x_n + x_n \frac{s - x_n^3}{s + 2x_n^3}$$

from which we see that $x_{n+1} \ge x_n$, since $s \ge x_n^3$. Thus, for lower initialization, we have $1 \le x_0 \le x_1 \le x_2 \le \cdots \le x_n \le \sqrt[3]{s}$, or $x_n \ge 1$ for all n. For upper initialization, we clearly have $x_n \ge \sqrt[3]{s} \ge 1$, so that in both cases $x_n \ge 1$ for all n. Finally, $x_n \ge 1$ and $\sqrt[3]{s} \ge 1$ imply

$$\frac{x_n + \sqrt[3]{s}}{2x_n^3 + s} \le \frac{x_n + s}{2x_n^3 + s} \le \frac{2x_n + s}{2x_n^3 + s} \le \frac{2x_n^3 + s}{2x_n^3 + s} = 1$$

2.21 The second-order iterations for $\sqrt{s}$ satisfy (2.13), that is,

$$x_{n+1} - \sqrt{s} = \frac{1}{2x_n} \left(x_n - \sqrt{s}\right)^2$$

whereas the corresponding iterations for $\sqrt[3]{s}$ satisfy (2.53), that is,

$$x_{n+1} - \sqrt[3]{s} = \left(\frac{2}{3x_n} + \frac{\sqrt[3]{s}}{3x_n^2}\right) \left(x_n - \sqrt[3]{s}\right)^2$$

In both cases, because always $x_0 \ge 1 > 0$, the right-hand sides are nonnegative, and hence so are the left-hand sides. Consequently, the approximations $x_1, x_2, \ldots$ are from above.

2.22 The third-order iterations for $\sqrt{s}$ satisfy (2.39), that is,

$$x_{n+1} - \sqrt{s} = \frac{1}{3x_n^2 + s} \left(x_n - \sqrt{s}\right)^3$$

Because $1/(3x_n^2 + s)$ is positive, it is clear that $x_0 \ge \sqrt{s}$ implies $x_n \ge \sqrt{s}$ for all n; that is, we have approximations from above. On the other hand, $x_0 \le \sqrt{s}$ implies $x_n \le \sqrt{s}$ for all n, and we have approximations from below. For $\sqrt[3]{s}$, we have seen in Exercise 2.19 that

$$x_{n+1} - \sqrt[3]{s} = \frac{x_n + \sqrt[3]{s}}{2x_n^3 + s} \left(x_n - \sqrt[3]{s}\right)^3$$

with $0 < \left(x_n + \sqrt[3]{s}\right) / \left(2x_n^3 + s\right) \le 1$. Now we can employ the same reasoning used for the square-root approximations.

2.23 For $\sqrt[5]{s}$, we represent any positive number d in the form

$$d = s \cdot 32^m = s \cdot 2^{5m}\,, \qquad 1 \le s < 32$$

with m an appropriate integer. Thus, $\sqrt[5]{d} = \sqrt[5]{s} \cdot 2^m$, where $1 \le \sqrt[5]{s} < 2$. Now, generalizing the approach in (2.10) and (2.48), we set

$$x_{n+1} - \sqrt[5]{s} = \frac{\left(x_n - \sqrt[5]{s}\right)^2}{\alpha_n} + \frac{\left(x_n - \sqrt[5]{s}\right)^3}{\beta_n} + \frac{\left(x_n - \sqrt[5]{s}\right)^4}{\gamma_n} + \frac{\left(x_n - \sqrt[5]{s}\right)^5}{\delta_n}$$

Now, express x_{n+1} is ascending powers of $\sqrt[5]{s}$, and require the vanishing of the coefficients of $\sqrt[5]{s}$, $\left(\sqrt[5]{s}\right)^2$, $\left(\sqrt[5]{s}\right)^3$, and $\left(\sqrt[5]{s}\right)^4$. This procedure yields

$$\alpha_n = \frac{x_n}{2}, \qquad \beta_n = \frac{-x_n^2}{2}, \qquad \gamma_n = x_n^3, \qquad \delta_n = -5x_n^4$$

which in turn leads to the iterative formula

$$x_{n+1} = \frac{4}{5}\left(x_n + \frac{\frac{s}{4}}{x_n^4}\right)$$

Moreover, using the values obtained for α_n, β_n, γ_n, and δ_n, we also have

$$x_{n+1} - \sqrt[5]{s} = \frac{4x_n^3 + 3\sqrt[5]{s}\, x_n^2 + 2\left(\sqrt[5]{s}\right)^2 x_n + \left(\sqrt[5]{s}\right)^3}{5x_n^4}\left(x_n - \sqrt[5]{s}\right)^2$$

This expression, together with $1 \le \sqrt[5]{s}$, shows that $1 \le \sqrt[5]{s} \le x_n$, for $n = 1, 2, 3, \ldots$. We next choose an upper initialization $x_0 \ge \sqrt[5]{s}$, and conclude from the preceding expression that

$$0 \le x_{n+1} - \sqrt[5]{s} \le \frac{2}{x_n}\left(x_n - \sqrt[5]{s}\right)^2 \le 2\left(x_n - \sqrt[5]{s}\right)^2$$

for all n. Using this inequality recursively as we did in (2.16)–(2.21), with $K = 2$, we obtain

$$0 \le x_{n+1} - \sqrt[5]{s} \le \frac{1}{2}\left[2(x_0 - \sqrt[5]{s}\,)\right]^{2^n} \le \frac{1}{2}\,(2\lambda)^{2^n}$$

where $0 \le x_0 - \sqrt[5]{s} \le \lambda$. If we use the simple upper initialization

$$x_0 = \begin{cases} 1.5, & 1 \le s < 7.59375 = (1.5)^5 \\ 2.0, & 7.59375 \le s < 32 \end{cases}$$

we are guaranteed that $0 \le x_0 - \sqrt[5]{s} \le \frac{1}{2} = \lambda$. However, because the error decay is proportional to $(2\lambda)^{2^n}$, $\lambda = \frac{1}{2}$ does not suffice. We therefore choose the upper initialization

$$x_0 = \begin{cases} 1.25, & 1.00 \le s < (1.25)^5 \\ 1.50, & (1.25)^5 \le s < (1.50)^5 \\ 1.75, & (1.50)^5 \le s < (1.75)^5 \\ 2.00, & (1.75)^5 \le s < 32 \end{cases}$$

which guarantees $0 \le x_0 - \sqrt[5]{s} \le \frac{1}{4} = \lambda$, and thus

$$0 \le x_n - \sqrt[5]{s} \le \frac{1}{2}\left(\frac{1}{2}\right)^{2^n}$$

If desired, an improved initialization may also be constructed along the lines of (2.58)–(2.69).

2.24 The lower initialization (2.86) gives $\lambda = \frac{1}{2}$. Substituting this value in (2.85), and using $K = \frac{3}{2}$ from (2.83), we find

$$|x_n - \sqrt{s}| \le \frac{2}{3}\left(\frac{3}{4}\right)^{2^n}$$

Now, for $n = 7$, we have

$$\frac{2}{3}\left(\frac{3}{4}\right)^{2^7} = \frac{2}{3}\cdot\left(\frac{3}{4}\right)^{128} < \frac{1}{2}\cdot 10^{-15}$$

which guarantees 15 correct figures.

2.25 The polynomial iterative formula of third order for $\sqrt{s}$, according to (2.87), is

$$\begin{aligned} x_{n+1} = {} & \alpha x_n^5 + (\beta - 3\alpha\sqrt{s})x_n^4 + (3\alpha s + \gamma - 3\beta\sqrt{s})x_n^3 \\ & + (3\beta s - 3\gamma\sqrt{s} - \alpha s\sqrt{s})x_n^2 \\ & + (3\gamma s - \beta s\sqrt{s})x_n + (1 - \gamma s)\sqrt{s} \end{aligned}$$

To eliminate the dependence on $\sqrt{s}$ in the last term, we must choose $\gamma = 1/s$, which in turn leaves us with

$$\begin{aligned} x_{n+1} = {} & \alpha x_n^5 + (\beta - 3\alpha\sqrt{s})x_n^4 + \frac{3\alpha s^2 + 1 - 3\beta s\sqrt{s}}{s}x_n^3 \\ & + \frac{3\beta s\sqrt{s} - 3 - \alpha s^2}{\sqrt{s}}x_n^2 + (3 - \beta s\sqrt{s})x_n \end{aligned}$$

Clearly α must not "contain" $\sqrt{s}$ because no coefficient of a power of x_n is allowed to contain $\sqrt{s}$. Accordingly, from the coefficient of x_n, we conclude that $\beta = \delta/\sqrt{s}$, where δ is "$\sqrt{s}$ free." Thus, we have

$$x_{n+1} = \alpha x_n^5 + \frac{\delta - 3\alpha s}{\sqrt{s}} x_n^4 + \frac{3\alpha s^2 - 3\delta s + 1}{s} x_n^3 + \frac{3\delta s - \alpha s^2 - 3}{\sqrt{s}} x_n^2 + (3 - \delta s) x_n$$

From the coefficient of x_n^4 we see that we must have $\delta = 3\alpha s$, so that

$$x_{n+1} = \alpha x_n^5 + \frac{1 - 6\alpha s^2}{s} x_n^3 + \frac{8\alpha s^2 - 3}{\sqrt{s}} x_n^2 + 3(1 - \alpha s^2) x_n$$

Again, from the coefficient of x_n^2 we conclude that $\alpha = 3/(8s^2)$, which is indeed $\sqrt{s}$ free. This in turn implies $\delta = 9/(8s)$ and $\beta = 9/(8s\sqrt{s})$. Those are indeed the values of α, β, and γ given by (2.88).

Remark: It would appear that the choice $\gamma = 1/s + Q(s)/\sqrt{s}$, where $Q(s)$ is $\sqrt{s}$ free, would equally well eliminate the dependence on $\sqrt{s}$ in the last term in the expression for x_{n+1}, as would the choice $\beta = \delta/\sqrt{s} + R(s)$, where both δ and $R(s)$ are $\sqrt{s}$ free. A tedious though elementary calculation shows, however, that both Q and R must be identically zero, so that the coefficients of all powers of x_n will indeed be $\sqrt{s}$ free.

2.26 The suggested polynomial iterative formula of second order for $\sqrt[3]{s}$ takes the form

$$x_{n+1} = \alpha x_n^4 + \left(\beta - 2\alpha\sqrt[3]{s}\right) x_n^3 + \left(\gamma + \alpha(\sqrt[3]{s})^2 - 2\beta\sqrt[3]{s}\right) x_n^2 + \left(\beta\sqrt[3]{s} - 2\gamma\right)\sqrt[3]{s}\, x_n + \left(\gamma\sqrt[3]{s} + 1\right)\sqrt[3]{s}$$

To eliminate the dependence on $\sqrt[3]{s}$ in the last term, we take $\gamma = -1/\sqrt[3]{s}$. In this connection, see the remark at the end of Exercise 2.25. Analogous considerations apply here as well. Having chosen γ, we are left with

$$x_{n+1} = \alpha x_n^4 + \left(\beta - 2\alpha\sqrt[3]{s}\right) x_n^3 + \frac{\alpha s - 2\beta(\sqrt[3]{s})^2 - 1}{\sqrt[3]{s}} x_n^2 + \left(\beta(\sqrt[3]{s})^2 + 2\right) x_n$$

Clearly α must be $\sqrt[3]{s}$ free. Accordingly, from the coefficient of x_n^3 we are led to take $\beta = 2\alpha\sqrt[3]{s}$, and so

$$x_{n+1} = \alpha x_n^4 - \frac{3\alpha s + 1}{\sqrt[3]{s}} x_n^2 + 2(1 + \alpha s)x_n$$

Finally, from the coefficient of x_n^2 we are led to $\alpha = -1/(3s)$, from which it follows that

$$x_{n+1} = \frac{-1}{3s} x_n^4 + \frac{4}{3} x_n = x_n \left(\frac{4}{3} - \frac{x_n^3}{3s} \right)$$

as required. Observe that (taking into account the earlier remark) the very fact that α, β, and γ were determined uniquely shows that two coefficients are not enough.

2.27 We substitute the values $\alpha = -1/(3s)$, $\beta = -(2/3)/\left(\sqrt[3]{s}\right)^2$, and $\gamma = -1/\sqrt[3]{s}$ found in Exercise 2.26 into

$$x_{n+1} - \sqrt[3]{s} = \left(\alpha x_n^2 + \beta x_n + \gamma\right)\left(x_n - \sqrt[3]{s}\right)^2$$

This yields

$$\frac{x_{n+1} - \sqrt[3]{s}}{\left(x_n - \sqrt[3]{s}\right)^2} = -\frac{\left(x_n + \sqrt[3]{s}\right)^2 + 2\left(\sqrt[3]{s}\right)^2}{3s}$$

from which we deduce that $x_k \leq \sqrt[3]{s}$ for all $k = 1, 2, \ldots$. Because we are assuming the lower initialization $x_0 = (s+6)/7$, we find (see Fig. 2.3) that $x_0 \leq \sqrt[3]{s}$, too. Accordingly, we write

$$\frac{|x_{n+1} - \sqrt[3]{s}|}{|x_n - \sqrt[3]{s}|^2} = \frac{\left(x_n + \sqrt[3]{s}\right)^2 + 2\left(\sqrt[3]{s}\right)^2}{3s} \leq \frac{\left(2\sqrt[3]{s}\right)^2 + 2\left(\sqrt[3]{s}\right)^2}{3s}$$
$$= \frac{2}{\sqrt[3]{s}} \leq 2$$

since $1 \leq \sqrt[3]{s} < 2$, establishing the required result

$$|x_{n+1} - \sqrt[3]{s}| \leq 2|x_n - \sqrt[3]{s}|^2$$

Now suppose that our lower initialization $x_0 = (s+6)/7$ satisfies $0 \leq \sqrt[3]{s} - x_0 \leq \delta$, or $|x_0 - \sqrt[3]{s}| \leq \delta$. Then,

$$|x_1 - \sqrt[3]{s}| \leq 2|x_0 - \sqrt[3]{s}|^2 \leq 2\delta^2$$
$$|x_2 - \sqrt[3]{s}| \leq 2|x_1 - \sqrt[3]{s}|^2 \leq 2(2\delta^2)^2 = 2^3\delta^4$$
$$|x_3 - \sqrt[3]{s}| \leq 2|x_2 - \sqrt[3]{s}|^2 \leq 2(2^3\delta^4)^2 = 2^7\delta^8$$

and so on, and generally,

$$|x_n - \sqrt[3]{s}| \leq 2^{2^n - 1}\delta^{2^n} = \frac{1}{2}(2\delta)^{2^n}$$

as required. Finally, because for our lower initialization [see inequality (2.70) and Fig. 2.3] we have $|x_0 - \sqrt[3]{s}| \leq \frac{1}{6} = \delta$, we find that $|x_n - \sqrt[3]{s}| \leq (\frac{1}{2})/3^{2^n}$

2.30 We have

$$f_n = x_n^2 - s$$
$$f_n' = 2x_n$$
$$f_n'' = 2$$

Consequently,

$$\begin{aligned} x_{n+1} &= x_n - \frac{f_n}{f_n'} - \frac{f_n^2 f_n''}{2(f_n')^3} \\ &= \frac{3x_n}{8} + \frac{3s}{4x_n} - \frac{s^2}{8x_n^3} \end{aligned}$$

Next, we subtract $\sqrt{s}$ from both sides and arrive at

$$x_{n+1} - \sqrt{s} = \frac{3x_n^4 - 8\sqrt{s}x_n^3 + 6sx_n^2 - s^2}{8x_n^3}$$

Because we are dealing with a third-order method, it is clear that the numerator is divisible by $(x_n - \sqrt{s})^3$. Using long division we find

$$\frac{3x_n^4 - 8\sqrt{s}x_n^3 + 6sx_n^2 - s^2}{x_n^3 - 3\sqrt{s}x_n^2 + 3sx_n - s\sqrt{s}} = 3x_n + \sqrt{s}$$

Thus,

$$x_{n+1} - \sqrt{s} = \frac{3x_n + \sqrt{s}}{8x_n^3}(x_n - \sqrt{s})^3$$

Now,

$$x_{n+1} - \sqrt{s} = \left(\frac{3}{8x_n^2} + \frac{\sqrt{s}}{8x_n^3}\right)(x_n - \sqrt{s})^3$$

from which it follows that if $x_0 \geq \sqrt{s}$, then also $x_n \geq \sqrt{s}$ for $n = 0, 1, 2, \ldots$. Consequently,

$$\begin{aligned} |x_{n+1} - \sqrt{s}| &\leq \left[\frac{3}{8(\sqrt{s})^2} + \frac{\sqrt{s}}{8(\sqrt{s})^3}\right] \cdot |x_n - \sqrt{s}|^3 \\ &\leq \frac{1}{2s}|x_n - \sqrt{s}|^3 \leq \frac{1}{2}|x_n - \sqrt{s}|^3 \end{aligned}$$

since $1 \leq s < 4$.

2.31 We have

$$f_n = 1 - \frac{s}{x_n^2}$$
$$f_n' = \frac{2s}{x_n^3}$$
$$f_n'' = \frac{-6s}{x_n^4}$$

Using the Newton–Raphson method (2.100), we obtain

$$x_{n+1} = x_n - \frac{f_n}{f_n'} = x_n\left(\frac{3}{2} - \frac{x_n^2}{2s}\right)$$

Moreover, using the extended Newton method (2.102), we find

$$\begin{aligned} x_{n+1} &= x_n - \frac{f_n}{f_n'} - \frac{f_n^2 f_n''}{2(f_n')^3} \\ &= x_n\left(\frac{3}{2} - \frac{x_n^2}{2s}\right) + \frac{(1 - s/x_n^2)^2 \cdot 6s/x_n^4}{16s^3/x_n^9} \\ &= \frac{x_n}{8s^2}\left(3x_n^4 - 10sx_n^2 + 15s^2\right) \end{aligned}$$

which is equivalent to (2.89).

2.32 We have

$$f_n = x_n^3 - s$$
$$f_n' = 3x_n^2$$

Thus,

$$\begin{aligned} x_{n+1} &= x_n - \frac{f_n}{f_n'} = x_n - \frac{x_n^3 - s}{3x_n^2} \\ &= \frac{2}{3}\left(x_n + \frac{\frac{s}{2}}{x_n^2}\right) \end{aligned}$$

2.33 We have

$$f_n = 1 - \frac{s}{x_n^k}$$
$$f_n' = \frac{ks}{x_n^{k+1}}$$
$$f_n'' = \frac{-k(k+1)s}{x_n^{k+2}}$$

Using the Newton–Raphson method (2.100), we obtain

$$x_{n+1} = x_n - \frac{1 - s/x_n^k}{ks/x_n^{k+1}} = x_n\left(\frac{k+1}{k} - \frac{x_n^k}{ks}\right)$$

which reduces to the result of Exercise 2.26 for $k = 3$. Moreover, using the extended Newton method (2.102), we find

$$\begin{aligned} x_{n+1} &= x_n \left(\frac{k+1}{k} - \frac{x_n^k}{ks} \right) + \frac{(1 - s/x_n^k)^2 k(k+1)s/x_n^{k+2}}{2k^3 s^3 / x_n^{3k+3}} \\ &= x_n \left[\frac{k+1}{2k^2 s^2} x_n^{2k} - \frac{2k+1}{k^2 s}\, x_n^k + \frac{(k+1)(2k+1)}{2k^2} \right] \end{aligned}$$

Chapter 3

3.1 Because the function $f(x)$ is increasing, we obtain a result that is analogous to (3.1) in the form

$$L_n = \sum_{j=1}^{n} hf(x_{j-1}) \le S \le \sum_{j=1}^{n} hf(x_j) = R_n$$

as can be seen from figures analogous to Figs. 3.1 and 3.2, appropriate to an increasing $f(x)$. Clearly, the difference $|R_n - L_n|$ is again a bound for the global errors $|S - L_n|$ and $|S - R_n|$ and is given by (3.2). Constructing figures analogous to Figs. 3.3 and 3.4, appropriate to an increasing $f(x)$, we see the geometric interpretation of this bound, and in this case the global error-bound column is positioned at the left end of the interval $[a, b]$.

3.4 For this case $x_j = jh$, so that $f(x_j) = (jh)^2$, and we have

$$R_n = \sum_{j=1}^{n} h(jh)^2 = h^3 \sum_{j=1}^{n} j^2$$

$$\begin{aligned} L_n &= \sum_{j=1}^{n} h\big[(j-1)h\big]^2 = h^3 \sum_{j=1}^{n} (j-1)^2 = h^3 \sum_{i=0}^{n-1} i^2 \\ &= h^3 \sum_{i=1}^{n-1} i^2 \end{aligned}$$

where we have changed the summation index from j to i. We now recall the formula

$$1^2 + 2^2 + 3^2 + \cdots + m^2 = \frac{m(m+1)(2m+1)}{6}$$

which is a standard exercise in the study of mathematical

induction. Using this formula with $m = n$ for R_n and with $m = n-1$ for L_n, we reach

$$h^3\frac{(n-1)n(2n-1)}{6} \leq S \leq h^3\frac{n(n+1)(2n+1)}{6}$$

Because $h = b/n$ for this case ($a = 0$), it follows that

$$\frac{b^3}{3}\left(1-\frac{1}{n}\right)\left(1-\frac{1}{2n}\right) \leq S \leq \frac{b^3}{3}\left(1+\frac{1}{n}\right)\left(1+\frac{1}{2n}\right)$$

or

$$\frac{b^3}{3}\left(1-\frac{\frac{3}{2}}{n}+\frac{1}{2n^2}\right) \leq S \leq \frac{b^3}{3}\left(1+\frac{\frac{3}{2}}{n}+\frac{1}{2n^2}\right)$$

as required. Note that the error bound B in (3.2), given by the difference between the extreme sides of the inequality, is b^3/n. Clearly, for large values of n we see that S approaches $b^3/3$.

3.7 Using the rectangular method in $[a, c]$, where $f(x)$ is increasing we have

$$L_n \leq S_a^c \leq R_n$$

where S_a^c denotes the area under the graph of $f(x)$ from $x = a$ to $x = c$. Similarly, for the decreasing part of $f(x)$ we have

$$\widetilde{R}_m \leq S_c^b \leq \widetilde{L}_m$$

where $\widetilde{R}_m$ and $\widetilde{L}_m$ are the appropriate sums for the interval $[c, b]$. Finally,

$$L_n + \widetilde{R}_m \leq S_a^b \leq R_n + \widetilde{L}_m$$

3.8 In this case, K of (3.16) is given by

$$K = \frac{(3-0)^2}{2}\left|\frac{(3+h)^2-3^2}{h} - \frac{(0+h)^2-0^2}{h}\right|$$

$$= \frac{9}{2}|(6+h)-h| = 27 = D$$

The choice $K = D$ is permitted here because K turns out to be independent of h (and n). To guarantee q correct figures we must require that $27/n^2 \leq (1/2)\cdot 10^{-q}$, that is, $n \geq \sqrt{54}\cdot 10^{q/2}$. For $q = 4$, $n = 735$ will suffice, whereas for $q = 5$, $n = 2324$ will do.

3.9 Here $f(x) = x^3$, $a = \frac{1}{2}$, and $b = \frac{3}{2}$. Consequently, the error bound (3.2) for the rectangular method is given by

$$B = \frac{\left[\left(\frac{3}{2}\right)^3 - \left(\frac{1}{2}\right)^3\right]\left[\frac{3}{2} - \frac{1}{2}\right]}{n} = \frac{\frac{13}{4}}{n}$$

whereas the trapezoidal error bound is found from (3.16) to be

$$\frac{K}{n^2} = \frac{\left(\frac{3}{2} - \frac{1}{2}\right)^2}{2n^2}\left|\frac{\left(\frac{3}{2}+h\right)^3 - \left(\frac{3}{2}\right)^3}{h} - \frac{\left(\frac{1}{2}+h\right)^3 - \left(\frac{1}{2}\right)^3}{h}\right|$$

$$= \frac{6+3h}{2n^2} \le \frac{\frac{9}{2}}{n^2} = \frac{D}{n^2}$$

since clearly h cannot exceed 1. Requiring four correct figures, we must have $(13/4)/n \le (1/2) \cdot 10^{-4}$ and $(9/2)/n^2 \le (1/2) \cdot 10^{-4}$, respectively, so that $n \ge 65{,}000$ in the first case and $n \ge 300$ in the second.

3.10 In this case $f(x) = \sqrt{1+x^2}$, $a = 1$, and $b = 3$, and (3.16) yields

$$K = \frac{(3-1)^2}{2}\left|\frac{f(3+h) - f(3)}{h} - \frac{f(1+h) - f(1)}{h}\right|$$

$$< \frac{4}{2} \cdot \frac{f(3+h) - f(3)}{h} = 2\frac{\sqrt{1+(3+h)^2} - \sqrt{10}}{h}$$

$$= 2\frac{(\sqrt{1+(3+h)^2} - \sqrt{10})(\sqrt{1+(3+h)^2} + \sqrt{10})}{h \cdot (\sqrt{1+(3+h)^2} + \sqrt{10})}$$

$$= \frac{12+2h}{\sqrt{1+(3+h)^2} + \sqrt{10}} \le \frac{12+2h}{2\sqrt{10}} < \frac{8}{\sqrt{10}}$$

since h cannot exceed 2. Now, requiring $(8/\sqrt{10})/n^2 \le (1/2) \cdot 10^{-5}$ so that $n \ge 400 \cdot \sqrt[4]{10}$, we find that $n = 712$ suffices.

3.11 From Fig. 3.10 we see that the area under the graph of a convex function in a typical strip is sandwiched between the area of the midpoint trapezoid and that of the upper trapezoid. This being true for each strip, we sum over all strips to obtain

$$|S - T_n| + |S - M_n| = |T_n - M_n|$$

Deleting either one of the two terms on the left-hand side clearly decreases its value, leading to

$$|S - T_n| \leq |T_n - M_n|, \qquad |S - M_n| \leq |T_n - M_n|$$

The argument for a concave function is entirely analogous.

3.12 Let us examine an enlarged drawing of a typical strip in which we approximate the area under the graph of $f(x) = 1/x$ by the trapezoidal method as well as by the midpoint rule. The local error for the trapezoidal method is estimated by the area of a triangle whose base is the extension of the roof of the neighboring trapezoid (see Fig. 3.7). On the other hand, the local error for the midpoint rule is estimated by the area of a "thin" trapezoid, generated by the difference between the trapezoids of the midpoint rule and the trapezoidal method.

3.14 In this case, we have to compute the global error bound given by

$$\begin{aligned}\sum_{j=1}^{n} \frac{3}{4}\left[\left(j - \frac{1}{2}\right)h\right]h^3 &= \frac{3}{4}h^4 \sum_{j=1}^{n}\left(j - \frac{1}{2}\right) \\ &= \frac{3}{4}h^4\left[\frac{1}{2} + \frac{3}{2} + \frac{5}{2} + \cdots + \left(n - \frac{1}{2}\right)\right] \\ &= \frac{3}{4}h^4 \frac{\frac{1}{2} + \left(n - \frac{1}{2}\right)}{2} n = \frac{3}{8}n^2h^4 = \frac{6}{n^2}\end{aligned}$$

since $h = 2/n$. Finally we have

$$|S - M_n| \leq \frac{6}{n^2}$$

which is twice as sharp as (3.30).

3.15 Here we have, in analogy to (3.28),

$$e_j \leq \frac{3}{4}\left[a + \left(j - \frac{1}{2}\right)h\right]h^3$$

where $j = 1, 2, \ldots, n$. Using the coarse estimate $a + (j - \frac{1}{2})h < b$ for all j, we obtain

$$\sum_{j=1}^{n} e_j < \sum_{j=1}^{n} \frac{3}{4}bh^3 = \frac{3}{4}bh^3n = \frac{3(b-a)^3}{4n^2}b$$

since $h = (b-a)/n$. On the other hand, if we repeat Exercise 3.14, we get

$$\sum_{j=1}^{n} e_j \le \sum_{j=1}^{n} \frac{3}{4}\left[a + \left(j - \frac{1}{2}\right)h\right] h^3 = \frac{3}{4}h^3 \sum_{j=1}^{n}\left[a + \left(j - \frac{1}{2}\right)h\right]$$
$$= \frac{3}{4}h^3\left[\left(a + \frac{h}{2}\right) + \left(a + \frac{3h}{2}\right) + \cdots + \left(b - \frac{h}{2}\right)\right]$$

since $a + (n - \frac{1}{2})h = b - \frac{h}{2}$. Summing this arithmetic progression, we reach

$$\sum_{j=1}^{n} e_j \le \frac{3}{4}h^3 \frac{\left(a + \frac{h}{2}\right) + \left(b - \frac{h}{2}\right)}{2} n = \frac{3(b-a)^3}{4n^2} \cdot \frac{a+b}{2}$$

which is clearly sharper since $(a+b)/2 < b$.

3.16 In this case, $f(x) = 1/x^2$, $a = 2$, and $b = 5$. By (3.22), the local error of the midpoint rule is given by

$$e_j \le \frac{h}{2}\left[\frac{1}{(u - h/2)^2} + \frac{1}{(u + h/2)^2} - \frac{2}{u^2}\right]$$

where $u = x_{j-\frac{1}{2}} = 2 + (j - \frac{1}{2})h$, $1 \le j \le n$. Carrying out the indicated operations, we reach

$$e_j \le \frac{h}{2} \frac{(12u^2h^2 - h^4)/8}{u^2(u - h/2)^2(u + h/2)^2} < \frac{3h^3}{4(u - h/2)^2(u + h/2)^2}$$
$$\le \frac{3h^3}{4(u - h/2)^4} \le \frac{3h^3}{4a^4} = \frac{3h^3}{64}$$

Thus the global error bound is given by

$$\sum_{j=1}^{n} e_j \le n\frac{3h^3}{64} = \frac{\frac{81}{64}}{n^2}$$

To ensure six correct figures, we require $81/(64n^2) \le (1/2) \cdot 10^{-6}$, or $n \ge 1000 \cdot \sqrt{81/32}$, so that $n = 1591$ will do. On the other hand, for the rectangular method the global error bound (3.2) yields

$$\sum_{j=1}^{n} e_j \le \left|\frac{1}{5^2} - \frac{1}{2^2}\right| \frac{5-2}{n} = \frac{\frac{63}{100}}{n}$$

implying $n \ge 1.26 \cdot 10^6$ for six-digit accuracy.

3.17 In this case, $f(x) = \sqrt{x}$, $a = 1$, and $b = 4$. Because $\sqrt{x}$ is concave, the local error bound for the midpoint rule is

$$e_j \leq \frac{h}{2}\left[2\sqrt{u} - \sqrt{u - \frac{h}{2}} - \sqrt{u + \frac{h}{2}}\right]$$

where $u = x_{j-1/2} = 1 + (j - \frac{1}{2})h$. Now we write

$$e_j \leq \frac{h}{2}\left[\left(\sqrt{u} - \sqrt{u - \frac{h}{2}}\right) + \left(\sqrt{u} - \sqrt{u + \frac{h}{2}}\right)\right]$$

and multiply and divide each term by its algebraic conjugate. Thus, we reach

$$\begin{aligned} e_j &\leq \frac{h^2}{4}\left[\frac{1}{\sqrt{u} + \sqrt{u - h/2}} - \frac{1}{\sqrt{u} + \sqrt{u + h/2}}\right] \\ &= \frac{h^2}{4}\frac{\sqrt{u + h/2} - \sqrt{u - h/2}}{(\sqrt{u} + \sqrt{u - h/2})(\sqrt{u} + \sqrt{u + h/2})} \end{aligned}$$

Next, we multiply and divide by the algebraic conjugate of the numerator and arrive at

$$\begin{aligned} e_j &\leq \frac{h^3/4}{(\sqrt{u} + \sqrt{u - h/2})(\sqrt{u} + \sqrt{u + h/2})(\sqrt{u + h/2} + \sqrt{u - h/2})} \\ &\leq \frac{h^3/4}{(2\sqrt{u - h/2})^3} \leq \frac{h^3/4}{(2\sqrt{a})^3} = \frac{h^3}{32} \end{aligned}$$

Accordingly, the global error bound satisfies

$$\sum_{j=1}^{n} e_j \leq n\frac{h^3}{32} = \frac{\frac{27}{32}}{n^2}$$

so that $n = 1300$ suffices. On the other hand, for the rectangular method, the global error bound (3.2) yields

$$\sum_{j=1}^{n} e_j \leq |\sqrt{4} - \sqrt{1}|\frac{4-1}{n} = \frac{3}{n}$$

implying $n \geq 6 \cdot 10^6$ for six-digit accuracy.

3.19 In this case, we have

$$f(x) = \frac{1}{3+x^2}$$
$$f'(x) = \frac{-2x}{(3+x^2)^2}$$
$$f''(x) = \frac{6x^2-6}{(3+x^2)^3}$$
$$f'''(x) = \frac{24x(3-x^2)}{(3+x^2)^4}$$

To find $M_1 = \text{Max}\,|f'(x)|$ for $0 \le x \le 3$, we must check the values of $|f'(0)|$ and $|f'(3)|$, that is, the values at the endpoints, as well as $|f'(1)|$, the value at the relevant extremal point. We find $|f'(0)| = 0$, $|f'(3)| = \frac{1}{24}$, and $|f'(1)| = \frac{1}{8}$, so that $M_1 = \frac{1}{8}$. Similarly, to find M_2 we must check $|f''(0)|$, $|f''(3)|$, and $|f''(\sqrt{3})|$. Again, we find $|f''(0)| = \frac{2}{9}$, $|f''(3)| = \frac{1}{36}$, and $|f''(\sqrt{3})| = \frac{1}{18}$, so that $M_2 = \frac{2}{9}$. Thus, the various global error bounds represented in (3.35)–(3.37) are $|I - L_n| \le \frac{9}{16}/n$, $|I - R_n| \le \frac{9}{16}/n$, $|I - T_n| \le \frac{1}{2}/n^2$, and $|I - M_n| \le \frac{1}{4}/n^2$.

3.20 Our underlying function is $f(x) = x^2e^{-x^2}$, with $a = 0$ and $b = 4$. We find

$$f'(x) = -2xe^{-x^2}(x^2-1)$$
$$f''(x) = 2e^{-x^2}(2x^4 - 5x^2 + 1)$$
$$f'''(x) = -4xe^{-x^2}(2x^4 - 9x^2 + 6)$$

To find M_1, we must compare $|f'(0)|$, $|f'(4)|$, $|f'(\sqrt{5+\sqrt{17}}/2)|$, and $|f'(\sqrt{5-\sqrt{17}}/2)|$. It turns out that the last value is the largest, with a value of 0.5872089. Accordingly, we take $M_1 = 0.6$ and require $4.8/n \le 0.5 \cdot 10^{-q}$, or $n \ge 9.6 \cdot 10^q$, so that the rectangular method yields q correct figures. To find M_2, we must compare $|f''(0)|$, $|f''(4)|$, $|f''(\sqrt{9+\sqrt{33}}/2)|$, and $|f''(\sqrt{9-\sqrt{33}}/2)|$. It turns out that $|f''(0)| = 2$ is the largest, and we have $M_2 = 2$. Thus, for the trapezoidal method we require $\frac{32}{3}/n^2 \le \frac{1}{2} \cdot 10^{-q}$, or $n \ge 4.7 \cdot 10^{q/2}$. For the midpoint rule we must have $\frac{16}{3}/n^2 \le \frac{1}{2} \cdot 10^{-q}$, or $n \ge 3.3 \cdot 10^{q/2}$, to ensure q correct figures.

Chapter 4

4.1 We multiply the first row of the augmented table (4.3) of the pharmacist problem by (-4) and add it to the second, then multiply the first row by (-0.2) and add it to the third. This yields

0.1	0.3	0.5	100
0	–1.0	–1.7	–236
0	0.03	0.05	7

Next, we multiply the second row by 0.05 and add it to the third, to get the desired triangular form

0.1	0.3	0.5	100
0	–1.0	–1.7	–236
0	0	–0.001	–0.08

4.2 After operating with the first row on the three rows below it, we obtain

4	2	1	1	–3
0	8	1	2	12
0	0	3.5	0.5	7.5
0	0	0	7	7

Note that in this case, triangularization was achieved by operating merely with the first row. Needless to say, this is not a typical case.

4.3 Starting from the last row of the triangular form we obtained in Exercise 4.1, we find $X(3) = 80$. Substituting this value backward in the second row, we obtain $X(2) = 100$. Substituting farther backward in the first row, we reach $X(1) = 300$. You can check these results by substituting them into the original system (4.2).

4.4 Back substitution in the triangular form obtained in Exercise 4.2 yields, consecutively, $X(4) = 1$, $X(3) = 2$, $X(2) = 1$, and $X(1) = -2$.

4.5 Multiplying the last row of the augmented table (4.3) by (-2) and adding it to the second, then multiplying the last row by $(-\frac{10}{3})$ and adding it to the first, we obtain

1/30	0	0	10
0.36	0.02	0	110
0.02	0.09	0.15	27

Note that in this case there is no need to operate with the second row on the row above it (the first) because the operations with the third row yielded complete triangularization immediately. From the first row we find $X(1) = 300$. Forward substitution of $X(1)$ in the second row yields $X(2) = 100$, and further forward substitution in the third (last) row produces $X(3) = 80$.

4.6 Multiplying the last row of the table in Exercise 4.2 by $(-\frac{1}{6})$ and adding it to the third row, then multiplying the last row by $(-\frac{2}{3})$ and adding it to the second, then multiplying the last row by $(-\frac{1}{6})$ and adding it to the first row, we obtain

28/6	14/6	7/6	0	–28/6
64/6	80/6	22/6	0	–4/6
16/6	8/6	25/6	0	26/6
–4	–2	–1	6	10

We prefer to work with simple fractions to avoid round-off errors, and even more because this exercise is to be carried out by hand. Our next step is to operate with the third row on the two rows above it. To this end, we multiply the third row by $(-22/25)$ and add it to the second, then multiply the third row by $(-7/25)$ and add it to the first. This yields

98/25	49/25	0	0	–147/25
208/25	304/25	0	0	–112/25
16/6	8/6	25/6	0	26/6
–4	–2	–1	6	10

Finally, we multiply the second row by $(-49/304)$ and add it to the first row, to complete the triangularization in the form

49/19	0	0	0	−98/19
208/25	304/25	0	0	−112/25
16/6	8/6	25/6	0	26/6
−4	−2	−1	6	10

Now, clearly, the first row yields $X(1) = -2$. Forward substitution yields, consecutively, $X(2) = 1$, $X(3) = 2$, and $X(4) = 1$, in agreement with the solution obtained in Exercise 4.4.

4.7 Because the algorithm gradually produces zeros everywhere under the diagonal of the coefficient table, and because – whenever a new zero is created – the old zeros do not come into play at all, and because – as a result – the subdiagonal elements do not enter the backward substitution process, we can ignore the subdiagonal elements inside the innermost loop. In other words, we leave the subdiagonal elements unchanged, which means, in effect, that the index C of the innermost loop can start at the value R rather than at 1. Actually, this reasoning shows that C can even start at the value $R + 1$, thus increasing the computational efficiency.

4.11 After running the program and finding the solution column, divide the first row of the table by 30, the second row by 12, the third by 57, and the last by 43. You obtain

1	0.200	0.100	0.033	3
0.083	1	0.083	0.166	6
0.035	0.088	1	0.053	1
0.093	0.070	0.140	1	2

We find that the diagonal elements are *dominant*, so our system is "close" to the system

1	0	0	0	3
0	1	0	0	6
0	0	1	0	1
0	0	0	1	2

whose solution is given by the rightmost column. Thus, we are not surprised that the solution column of the given system is close to $X(1) = 3$, $X(2) = 6$, $X(3) = 1$, and $X(4) = 2$.

4.14 Based on Exercise 4.7, we can conclude that the first row – except its first element – operates on the $(N-1)$ rows below it, in each of which N elements are operated on (that is to say, the first element is excluded but the last element, belonging to the right-hand column, is included). Thus, the operations of the first row altogether consume $(N-1)N$ computational units, each of which consists of one multiplication and one addition. Analogously, the second row – except its first *two* elements – operates on the $(N-2)$ rows below it, in each of which $(N-1)$ elements are operated on. Eventually, the $(N-1)$ row operates on the Nth row, consuming $1 \cdot 2$ computational units. Thus, the total number of computational units, that is, the relevant computational complexity C is given by

$$\begin{aligned} C &= (N-1)N + (N-2)(N-1) + (N-3)(N-2) \\ &\quad + \cdots + 2 \cdot 3 + 1 \cdot 2 \\ &= [N^2 - N] + \left[(N-1)^2 - (N-1)\right] \\ &\quad + \left[(N-2)^2 - (N-2)\right] \\ &\quad + \cdots + [3^2 - 3] + [2^2 - 2] \end{aligned}$$

If we now add to C the vanishing quantity $[1^2 - 1]$, we see that

$$C = \sum_{j=1}^{N} j^2 - \sum_{j=1}^{N} j = \frac{N(N+1)(2N+1)}{6} - \frac{N(N+1)}{2}$$
$$= \frac{N(N+1)}{2}\left(\frac{2}{3}N + \frac{1}{3} - 1\right) = \frac{1}{3}N(N+1)(N-1)$$
$$= \frac{N^3}{3} - \frac{N}{3}$$

Therefore the triangularization has a computational complexity of (essentially) $N^3/3$ rather than $N^3/2$.

4.15 The diagonalization algorithm starts with the upper triangularization algorithm at the end of Section 4.5, followed by a simplified version of a lower triangularization (see Exercise 4.5). This simplification results from the fact that the only operating elements are the elements in the right hand column. This part of the algorithm can take the following form:

17. Repeat Steps 17–22 for R decreasing from N to 2.
18. Repeat Steps 18–21 for K decreasing from $K = R - 1$ to $K = 1$.
19. Compute $F = -T(K, R)/T(R, R)$.
20. Replace $T(K, N+1)$ by $[T(K, N+1) + F \cdot T(R, N+1)]$.
21. Close the loop started in Step 18.
22. Close the loop started in Step 17.
23. Repeat Steps 23–26 for R increasing from 1 to N.
24. Compute $X(R) = T(R, N+1)/T(R, R)$.
25. Print out the values of R and $X(R)$.
26. Close the loop started in Step 23.
27. End.

Note that in this part of the algorithm there is no need for the innermost column loop and that the reasoning behind the answer to Exercise 4.7 has been invoked.

4.16 The computational complexity of triangularization was shown in Exercise 4.14 to be (essentially) $N^3/3$. Had we followed the upper triangularization by a (nonsimplified, see Exercise 4.15) lower triangularization to diagonalize the coefficient table, the total complexity would have been $(2/3)N^3$. Comparing this with the computational complexity of triangularization followed by back substitution, which (see Sect. 4.5) is

$N^3/3 + N^2/2$, we see that diagonalization is less efficient and, indeed, twice as slow for large N.

4.17 The triangularization of the given table yields

4	2	1	1	3
0	0	–3	–2	2
0	0	16	14	–14
0	0	0	–7	7

The fourth row implies $X(4) = -1$, and the third row, in turn, shows that $X(3) = 0$. Substituting these values in the second row, we obtain

$$0 \cdot X(2) - 3 \cdot 0 - 2 \cdot (-1) = 2$$

from which we conclude that $X(2)$ can be chosen arbitrarily, say $X(2) = p$. Finally, the first row yields $X(1) = 1 - p/2$. Observe that we could have chosen $X(1)$ arbitrarily, say $X(1) = p$, in which case the first row would have implied $X(2) = 2 - 2p$.

4.18 The triangularization of the table yields

1	1	1	1	10
0	–3	1	–6	–27
0	0	0	0	0
0	0	0	0	0

A little reflection now shows that any two unknowns can be chosen arbitrarily, say $X(4) = p$ and $X(3) = q$, and that $X(2)$ and $X(1)$ can subsequently be determined from the first two rows in the form $X(2) = 9 - 2p + q/3$ and $X(1) = 1 + p - 4q/3$. If, instead of the number 17 in the rightmost column, there had been any other number, then the last element of the rightmost column in the triangularized table would have been nonzero. Thus, the fourth row is internally inconsistent, and so there is no solution.

4.19 Operating with the first row of (4.32) on the second, maintaining three-digit accuracy, we get

–1.41	2	0	1
0	0.01	1	1.71
0	2	–1.41	1

Pivoting, we interchange the second and third rows and then operate with the new second row on the third. This gives

–1.41	2	0	1
0	0.01	1	1.71
0	0	1.01	1.71

from which $X(3) = 1.69$, $X(2) = 1.70$, and $X(1) = 1.70$. This solution is clearly closer to the correct solution than the solution obtained without pivoting (see end of Sect. 4.6).

4.21 When this (in fact any) diagonally dominant system is solved, the pivoting procedure is never invoked, for the following reason. When the first row, for example, operates on the second row, it is multiplied by $-\frac{1}{30} \approx -0.0333$, and thus its addition to the second row hardly changes the latter, except for its first element, which becomes zero. Indeed, the second row becomes 0, 11.8, 0.9, 1.967, and the diagonal dominance is preserved. This being the case, no pivoting is called for at this stage. Moreover, the same reasoning applies to the operation of any row on the rows below it.

4.23 To begin with, we interchange the first two rows (pivoting), to obtain

20	40	2000	1840
0.1	–1	1	–1
8	15	200	135

Next we begin to triangularize this table, which leads us to

20	40	2000	1840
0	–1.2	–9	–10.2
0	–1	–600	–601

No further pivoting is necessary, and we can complete the triangularization to obtain

20	40	2000	1840
0	–1.2	–9	–10.2
0	0	–592.5	–592.5

Next we apply scaling to the original table, which yields

0.10	–1	1	–1
0.01	0.020	1	0.920
0.04	0.075	1	0.675

We see that no pivoting is necessary at this stage, so we begin to triangularize. Thus we obtain

0.1	–1	1	–1
0	0.120	0.9	1.020
0	0.475	0.6	1.075

Now we do have to apply pivoting, so we exchange the last two rows, complete the triangularization, and obtain

0.1	–1	1	–1
0	0.475	0.600	1.075
0	0	0.748	0.748

It is evident that scaling has indeed caused a different pivoting strategy.

4.24 The required 7×7 system is

0.1	–1	0	0	0	0	0	0.1
0	0.1	–1	0	0	0	0	–1
0	0	0.1	–1	0	0	0	0.1
0	0	0	0.1	–1	0	0	–1
0	0	0	0	0.1	–1	0	0.1
0	0	0	0	0	0.1	–1	–1
0	0	0	0	0	0	0.1	0.1

and its solution is given, obviously, by $X(7) = 1$, $X(6) = 0$, $X(5) = 1$, $X(4) = 0$, $X(3) = 1$, $X(2) = 0$, and $X(1) = 1$. Next we change $T(7, 8)$ from 0.1 to 0.101, leaving all other entries unchanged. Now the back substitution supplies $X(7) = 1.01$, $X(6) = 0.1$, $X(5) = 2$, $X(4) = 10$, $X(3) = 101$, $X(2) = 1000$, and $X(1) = 10001$. Advanced students are challenged to show that for an $n \times n$ system of this type, with n odd, the solution is given by $X(k) = 10^{n-2-k} + [1 - (-1)^k]/2$, $k = n, n-1, n-2, \ldots, 2, 1$.

4.26 Substituting the computed solution into the linear equation corresponding to (4.39) and subtracting from it the right-hand column, we obtain $R(1) = R(2) = R(3) = R(4) = 0$ and $R(5) = 0.001$. The error column, that is, the difference between the computed solution column, and the exact solution column, is found to be $E(5) = 0.01$, $E(4) = 0.1$, $E(3) = 1$, $E(2) = 10$, and $E(1) = 100$. The ratio of the maximal component of E to the maximal component of R is given by $\frac{100}{0.001} = 10^5$, and its huge size is characteristic of ill-conditioning. The corresponding calculations for the 7×7 system give

$$R(7) = 0.001,\ R(6) = R(5) = R(4) = R(3) = R(2) = R(1) = 0$$
$$E(7) = 0.01,\ E(6) = 0.1,\ E(5) = 1,\ E(4) = 10,\ E(3) = 100$$
$$E(2) = 1000,\ E(1) = 10000$$

and this time, the conditioning ratio is given by $\frac{10000}{0.001} = 10^7$ (!!!).

4.27 Triangularizing the system, we obtain (to five significant digits)

4.4	2.1	2.5	8.0
0	2.5636	0.70910	2.9091
0	0	2.5372	2.2553

from which, by back substitution, we get

$$X(3) = 0.88889, \qquad X(2) = 0.88891, \qquad X(1) = 0.88889$$

This yields

$$E(1) = 0.00001, \qquad E(2) = 0.00003, \qquad E(3) = 0.00001$$

and a short calculation gives $R(1) = R(2) = R(3) \approx 0.00003$. This time the conditioning ratio is approximately unity, and its small size is characteristic of well conditioning.

4.28 Let the system be scaled and then solved using pivoting. Every time pivoting is applied (with or without row exchange), the size (absolute value) of the pivot element should be tested. If this size turns out to be very much smaller than unity, say less than ε, then an ill-conditioning warning should be printed out. This testing process can be placed between Step 11 and Step 12 of the pivoting procedure introduced in Section 4.6. The size of this critical ε, in turn, is machine dependent.

Chapter 5

5.1 Consider first a regular hexagon inscribed in the unit circle and note that it is composed of six equilateral triangles with sides equal to unity. This hexagon's circumference is clearly equal to 6 and is less than the circle's circumference 2π. Hence $3 < \pi$. Next, note that the circumscribing hexagon is also composed of six equilateral triangles with height (which is also the median) equal to unity. Letting x equal the side of such a triangle, we have, by the theorem of Pythagoras, $1 + (x/2)^2 = x^2$, or $x = \frac{2}{3}\sqrt{3}$. The circumference of the circumscribing hexagon equals $6x = 4\sqrt{3}$ and is larger than the circle's circumference 2π. Hence $\pi < 2\sqrt{3}$.

5.2 In Fig. 5.2 we see that the triangle whose perpendicular sides are $a_n/2$ and b_n is similar to the triangle whose perpendicular sides are $c_n/2$ and unity. Because the corresponding sides of similar triangles are proportional, we obtain (5.5), that is,

$$\frac{\frac{c_n}{2}}{1} = \frac{\frac{a_n}{2}}{b_n}$$

Substituting for b_n from (5.3) we reach (5.6), that is,

$$c_n = \frac{a_n}{\sqrt{1 - \frac{a_n^2}{4}}}$$

and clearly the circumferences are given by nc_n and na_n. If we set $nc_n = C_n$ and $na_n = A_n$, we obtain

$$C_n = \frac{A_n}{\sqrt{1 - \frac{A_n^2}{4n^2}}}$$

Because A_n is bounded, we see that as n increases, the two circumferences tend to coalesce.

5.3 Because an inscribed n-gon generates an inscribed $2n$-gon by replacing each of its sides with the two sides of an isosceles triangle built on it and because the sum of the lengths of two sides of a triangle exceeds the length of the third side, it is clear that the sequence $3a_6, 6a_{12}, 12a_{24}, \ldots$ is increasing. Because the circumference of the inscribed n-gon "rounds out" toward the circle but obviously can never break out of it, the sequence is also bounded. For the same reason, the sequence $3c_6, 6c_{12}, 12c_{24}, \ldots$ associated with the circumscribed n-gons is bounded from below. Moreover, the cutting-corners process, by which a circumscribing n-gon generates a circumscribing $2n$-gon, actually replaces two sides of an isosceles triangle by its base, at each of the n corners. Thus, the sequence $3c_6, 6c_{12}, 24c_{24}, \ldots$ is decreasing.

5.4 Because both sides of the inequality $1/\sqrt{1 - u/4} \leq 1 + u/6$ are positive, we can square both sides and clear the fractions, to obtain

$$12u - 8u^2 - u^3 \geq 0$$

Now for $0 \leq u \leq 1$, we have $u^3 \leq u^2$ so that $-u^3 \geq -u^2$ and thus the inequality surely holds after we show that

$$12u - 8u^2 - u^2 = 12u - 9u^2 \geq 0$$

or

$$9u\left(\frac{4}{3}-u\right)\geq 0$$

which is clearly true for $0 \leq u \leq 1$. Note that the various steps are reversible, and thus the required inequality holds. Next we replace 6 by α in the inequality and reach

$$(8\alpha-\alpha^2)u-(2\alpha-4)u^2-u^3\geq 0$$

For $u = 0$ this inequality is evidently true, so we need only consider $0 < u \leq 1$. Dividing by u, we obtain

$$u^2+2(\alpha-2)u-\alpha(8-\alpha)\leq 0$$

We are looking for a value of α, where $6 \leq \alpha < 8$ (we know the inequality holds for $\alpha = 6$, and it cannot hold for $\alpha \geq 8$ because the sum of positive terms is positive). For any fixed α in this range and $0 < u \leq 1$, the left-hand side attains its maximal value for $u = 1$. Hence, setting $u = 1$, we find that α must satisfy

$$\alpha^2-6\alpha-3\leq 0$$

or

$$\left[\alpha-(3-2\sqrt{3})\right]\left[\alpha-(3+2\sqrt{3})\right]\leq 0$$

Therefore, α must satisfy $3-2\sqrt{3} \leq \alpha \leq 3+2\sqrt{3}$, and hence its maximal value is $\alpha = 3+2\sqrt{3} \approx 6.4641$. Here too the entire process is reversible, and therefore $\alpha = 3+2\sqrt{3}$ is the maximal value for which $1/\sqrt{1-u/4} \leq 1+u/\alpha$, for all $0 \leq u \leq 1$.

5.5 Starting with squares rather than hexagons, we have $a_4 = \sqrt{2}$ and $c_4 = 2$. Thus, (5.1) becomes

$$2.828 < 2\sqrt{2} = 2a_4 < \pi < 2c_4 = 4$$

which is clearly not as "tight" as (5.1). Equations (5.2)–(5.10) remain valid because they are independent of the initial configuration. On the other hand, (5.11) must clearly be modified into $na_n \leq nc_n \leq 8$, and thus (5.12) becomes

$$0\leq\frac{\pi-(n/2)a_n}{(n/2)a_n}\leq\frac{32/3}{n^2}$$

Hence (5.13) goes over into

$$\frac{\frac{32}{3}}{16\cdot 2^{2k}}\leq\frac{1}{2}\cdot 10^{-q}$$

because at present $n = 4 \cdot 2^k$ for $k = 0, 1, 2, \ldots$, starting with a square. Consequently, (5.14) turns into

$$\frac{2\sqrt{3}}{3} \cdot 10^{q/2} \leq 2^k$$

Now, since $10 < 2^{10/3}$ and $2\sqrt{3}/3 < 2^{2/9}$, it follows that if the number of Archimedes cycles satisfies $k \geq \frac{5}{3}q + \frac{2}{9}$, we are surely guaranteed q correct figures.

5.8 Clearly $\sqrt{1+h^2}+1 \geq 2$ and $\sqrt{3}+\sqrt{3-4h-4h^2} \leq 2\sqrt{3}$, for $0 \leq h \leq \frac{1}{2}$. Hence

$$\frac{2(1+h)}{\sqrt{3}+\sqrt{3-4h-4h^2}} \geq \frac{2(1+h)}{2\sqrt{3}} \quad \text{and} \quad \frac{h}{2} \geq \frac{h}{\sqrt{1+h^2}+1}$$

Thus, it will suffice to show that $(1+h)/\sqrt{3} > h/2$, or $1 + (1-\sqrt{3}/2)h > 0$, which is clearly valid. From Fig. 5.3 it can be seen that the secant line connecting the point $(1/2, \sqrt{3}/2)$ with a nearby point on the graph is steeper than the one connecting the point $(0, 1)$ with a nearby point on this graph. Because both secants have negative slopes, it follows that

$$\frac{\sqrt{1-h^2}-1}{h} > \frac{\sqrt{1-\left(\frac{1}{2}+h\right)^2}-\sqrt{1-\left(\frac{1}{2}\right)^2}}{h}$$

and by (5.18) and (5.19) this is equivalent to what we just proved.

5.10 The probability of getting the number 3 is $\frac{1}{6}$ because exactly one out of the die's six faces shows the number 3. The probability of getting an even number (2, 4, or 6) is $\frac{1}{2}$. Getting either 1 or 6 has probability $\frac{2}{6} = \frac{1}{3}$. The probability of getting the number 8 is clearly 0 because this will never happen, and similarly the probability of getting a whole number less than 7 is 1 because this is always the case. Finally, getting anything but 4 has probability $\frac{5}{6}$, or $1 - \frac{1}{6} = \frac{5}{6}$.

5.11 When we roll two fair dice, there are 36 possible outcomes: $1+1=2,\ 1+2=3,\ 2+1=3, \ldots,\ 5+6=11,\ 6+5=11,\ 6+6=12$. Therefore, we can get the outcome 4 in three ways ($1+3,\ 2+2$, and $3+1$), out of the total of 36 outcomes. Thus, the probability is $\frac{3}{36} = \frac{1}{12}$. The probability of getting the outcome 7 is $\frac{6}{36} = \frac{1}{6}$. Getting an outcome larger than 8 has the probability $(4+3+2+1)/36 = 5/18$.

When we roll three fair dice there are $6 \cdot 6 \cdot 6 = 216$ possible outcomes. The outcome 4 can come about through $1+1+2$, or $1+2+1$, or $2+1+1$ and hence has probability $\frac{3}{216} = \frac{1}{36}$. The outcome 7 can be obtained through $5+4+3+2+1 = 15$ distinct throws. For example, there are five possibilities in which one die (say the red one) shows the number 1 because there are five possibilities for the other two dice to show a total of 6: $1+5$, $2+4$, $3+3$, $4+2$, and $5+1$. Similarly, there are four possibilities in which the red die shows 2 because the other two dice can show a total of 5 in four different ways. Continuing in this manner, we find a total of 15 favorable outcomes, and hence the probability is $\frac{15}{216} = \frac{5}{72}$. To get the number of possibilities of obtaining an outcome larger than 8, we first find the number of possibilities of reaching an outcome of 8 or less, because this is easier. Now, the number of ways of getting the outcome 8 is $6+5+4+3+2+1 = 21$, which can be shown along the same lines we used to show how to get the outcome 7. Again, the same reasoning can be employed to show that the outcomes 6, 5, 4, and 3 can be obtained in 10, 6, 3, and 1 distinct ways, respectively. Obviously, we can never reach an outcome less than 3. Summing up, there are $21+15+10+6+3+1 = 56$ distinct ways of getting an outcome of 8 or less, and hence $216 - 56 = 160$ ways of obtaining an outcome larger than 8. Finally, the probability of reaching an outcome larger than 8, by rolling three dice, is $\frac{160}{216} = \frac{20}{27}$.

5.12 As defined in the text, $RAND(6)$ generates a random number r such that $0 \leq r < 6$, so that $(RAND(6)+1)$ is a random number x with $1 \leq x < 7$. Now, if $1 \leq x < 2$, then $INT(x) = 1$; if $2 \leq x < 3$, then $INT(x) = 2$; ... ; if $6 \leq x < 7$, then $INT(x) = 6$. Because x has an equal chance of falling anywhere in $[1, 7)$, the numbers 1, 2, 3, 4, 5, and 6 have equal chances of being generated by $INT(RAND(6) + 1)$, which therefore simulates a die.

5.13 To simulate two dice, we write $u = INT(RAND(6) + 1)$, $v = INT(RAND(6) + 1)$, and $w = u + v$. Note carefully that u and v simulate two *independent* rollings of a die, with independent outcomes, and hence w simulates the rolling of two dice.

5.18 Equation (5.33) shows that $\pi/4 = \arctan(1/2) + \arctan(1/3)$. Setting $u = 1/m$ and $v = 1/n$, we get from (5.32) that

$$\arctan\frac{1}{m} + \arctan\frac{1}{n} = \arctan\frac{1/m + 1/n}{1 - 1/(mn)} = \arctan\frac{n+m}{mn-1}$$

Choosing $m = 3,\ n = 7$, we find

$$\arctan\frac{10}{20} = \arctan\frac{1}{2} = \arctan\frac{1}{3} + \arctan\frac{1}{7}$$

and thus

$$\frac{\pi}{4} = 2\arctan\frac{1}{3} + \arctan\frac{1}{7}$$

Now we choose $m = 5,\ n = 8$, and find

$$\arctan\frac{13}{39} = \arctan\frac{1}{3} = \arctan\frac{1}{5} + \arctan\frac{1}{8}$$

Collecting the results, we see that

$$\frac{\pi}{4} = 2\arctan\frac{1}{5} + \arctan\frac{1}{7} + 2\arctan\frac{1}{8}$$

Next, choose $m = 8,\ n = 57$, and find

$$\arctan\frac{65}{455} = \arctan\frac{1}{7} = \arctan\frac{1}{8} + \arctan\frac{1}{57}$$

which combines with the preceding result to give

$$\frac{\pi}{4} = 2\arctan\frac{1}{5} + 3\arctan\frac{1}{8} + \arctan\frac{1}{57}$$

Now take $m = 7,\ n = 18$, and find

$$\arctan\frac{25}{125} = \arctan\frac{1}{5} = \arctan\frac{1}{7} + \arctan\frac{1}{18}$$

and thus, substituting also for $\arctan\frac{1}{7}$, we have

$$\frac{\pi}{4} = 5\arctan\frac{1}{8} + 2\arctan\frac{1}{18} + 3\arctan\frac{1}{57}$$

5.19 Working along the lines of (5.34) and (5.35), we can write

$$\begin{aligned}
&\left|2E_{1/5} + E_{1/7} + 2E_{1/8}\right| \\
&\quad\le \frac{2\left(\frac{1}{5}\right)^{2k+1} + \left(\frac{1}{7}\right)^{2k+1} + 2\left(\frac{1}{8}\right)^{2k+1}}{2k+1} \\
&\quad< \frac{5\left(\frac{1}{5}\right)^{2k+1}}{2k+1} < \frac{1}{2k}\left(\frac{1}{5}\right)^{2k} \le \frac{1}{2}\cdot 10^{-q}
\end{aligned}$$

so that q correct figures are ensured. This gives $10^q \leq k \cdot 5^{2k}$, so that $10^q \leq 5^{2k}$ will do. For example, $k \geq \frac{3}{4}q$ is more than satisfactory (compare with $k \geq \frac{5}{3}q$, following (5.35) in the text). A similar treatment of (5.37) yields

$$|6E_{1/8}+2E_{1/57}+E_{1/239}| \leq \frac{9\left(\frac{1}{8}\right)^{2k+1}}{2k+1} < \frac{1}{k}\left(\frac{1}{8}\right)^{2k} \leq \frac{1}{2}\cdot 10^{-q}$$

so that q correct figures are ensured. This gives $2 \cdot 10^q \leq k \cdot 8^{2k}$, so that $10^q \leq 8^{2k}$ will do, and in particular $k \geq \frac{5}{9}q$ is more than enough.

5.22 By (5.50), with $m \geq r$,

$$\binom{m}{r} = \frac{m!}{r!(m-r)!}$$

so that

$$\binom{m}{m-r} = \frac{m!}{(m-r)!\left[m-(m-r)\right]!}$$
$$= \frac{m!}{(m-r)!r!} = \binom{m}{r}$$

Moreover,

$$\binom{m}{r} = \frac{m!}{(m-r)!r!}$$
$$= \frac{1 \cdot 2 \cdot \cdots \cdot (m-r)(m-r+1)\cdots(m-1)m}{1 \cdot 2 \cdot \cdots \cdot (m-r) \cdot r!}$$
$$= \frac{m(m-1)\cdots(m-r+1)}{r!}$$

Next, for $r = 0$, we find from the above that

$$\binom{m}{0} = \binom{m}{m} = \frac{m!}{0!(m-0)!} = \frac{1}{0!} = 1$$

Moreover, for $r = 1$,

$$\binom{m}{1} = \binom{m}{m-1} = \frac{m!}{1!(m-1)!} = \frac{(m-1)!m}{(m-1)!} = m$$

5.23 From (5.53) we have, for example,

$$a_4 = 2 + \frac{1}{2!}\left(1 - \frac{1}{4}\right) + \frac{1}{3!}\left(1 - \frac{1}{4}\right)\left(1 - \frac{2}{4}\right) + \frac{1}{4!}\left(1 - \frac{1}{4}\right)\left(1 - \frac{2}{4}\right)\left(1 - \frac{3}{4}\right)$$

$$a_3 = 2 + \frac{1}{2!}\left(1 - \frac{1}{3}\right) + \frac{1}{3!}\left(1 - \frac{1}{3}\right)\left(1 - \frac{2}{3}\right)$$

and in general,

$$\begin{aligned} a_{k+1} = 2 &+ \frac{1}{2!}\left(1 - \frac{1}{k+1}\right) \\ &+ \frac{1}{3!}\left(1 - \frac{1}{k+1}\right)\left(1 - \frac{2}{k+1}\right) + \cdots \\ &+ \frac{1}{k!}\left(1 - \frac{1}{k+1}\right)\left(1 - \frac{2}{k+1}\right)\cdots\left(1 - \frac{k-1}{k+1}\right) \\ &+ \frac{1}{(k+1)!}\left(1 - \frac{1}{k+1}\right)\left(1 - \frac{2}{k+1}\right)\cdots\left(1 - \frac{k}{k+1}\right) \end{aligned}$$

$$\begin{aligned} a_k = 2 &+ \frac{1}{2!}\left(1 - \frac{1}{k}\right) + \frac{1}{3!}\left(1 - \frac{1}{k}\right)\left(1 - \frac{2}{k}\right) + \cdots \\ &+ \frac{1}{k!}\left(1 - \frac{1}{k}\right)\left(1 - \frac{2}{k}\right)\cdots\left(1 - \frac{k-1}{k}\right) \end{aligned}$$

We first note that every term of a_{k+1} and a_k is positive and that a_{k+1} has one term more than a_k. Let us now subtract a typical term of a_k – say, the one whose first factor is $1/r!$ – from the corresponding term of a_{k+1} having the same first factor. This difference equals

$$\frac{1}{r!}\left[\left(1 - \frac{1}{k+1}\right)\left(1 - \frac{2}{k+1}\right)\cdots\left(1 - \frac{r-1}{k+1}\right) - \left(1 - \frac{1}{k}\right)\left(1 - \frac{2}{k}\right)\cdots\left(1 - \frac{r-1}{k}\right)\right] = D_r$$

Because

$$\frac{j}{k+1} < \frac{j}{k}, \qquad j = 1, 2, \ldots, r-1$$

so that

$$-\frac{j}{k+1} > -\frac{j}{k}$$

we have

$$1 - \frac{j}{k+1} > 1 - \frac{j}{k}$$

and thus the first product in the square brackets is larger than the second. Thus, $D_r > 0$ for $r = 2, 3, \ldots$ and we find that

$$a_{k+1} - a_k = D_2 + D_3 + \cdots + D_k$$
$$+ \frac{1}{(k+1)!}\left(1 - \frac{1}{k+1}\right)\cdots\left(1 - \frac{k}{k+1}\right) > 0$$

that is, $a_{k+1} > a_k$. Since $a_1 = 2$, we also have $a_k \geq 2$.

5.24 We set $m = k+1,\ t_1 = 1,\ t_2 = t_3 = \cdots = t_{k+1} = 1 + 1/k$ in the arithmetic–geometric inequality and obtain

$$\left[1 \cdot \left(1 + \frac{1}{k}\right)^k\right]^{1/(k+1)} \leq \frac{1 + k\left(1 + \frac{1}{k}\right)}{k+1} = 1 + \frac{1}{k+1}$$

Raising both sides to the $(k+1)$ power yields

$$a_k = \left(1 + \frac{1}{k}\right)^k \leq \left(1 + \frac{1}{k+1}\right)^{k+1} = a_{k+1}$$

5.25 We set $m = k+2,\ t_1 = t_2 = \frac{1}{2},\ t_3 = t_4 = \cdots = t_{k+2} = 1 + 1/k$ in the arithmetic–geometric inequality and obtain

$$\left[\frac{1}{4}\left(1 + \frac{1}{k}\right)^k\right]^{1/(k+2)} \leq \frac{\frac{1}{2} + \frac{1}{2} + k\left(1 + \frac{1}{k}\right)}{k+2} = \frac{k+2}{k+2} = 1$$

Raising both sides to the $(k+2)$ power yields

$$a_k = \left(1 + \frac{1}{k}\right)^k \leq 4$$

Analogously, on setting $m = k+6,\ t_1 = t_2 = \cdots = t_6 = \frac{5}{6},$ $t_7 = t_8 = \cdots = t_{k+6} = 1 + \frac{1}{k}$, we are led to

$$\left[\left(\frac{5}{6}\right)^6 \left(1 + \frac{1}{k}\right)^k\right]^{1/(k+6)} \leq \frac{6 \cdot \frac{5}{6} + k\left(1 + \frac{1}{k}\right)}{k+6} = \frac{k+6}{k+6} = 1$$

and so,

$$\left(1 + \frac{1}{k}\right)^k \leq \left(\frac{6}{5}\right)^6 = (1.2)^6 < 2.986$$

Similarly, for $m = k + 11$, $t_1 = t_2 = \cdots = t_{11} = \frac{10}{11}$ and $t_{12} = t_{13} = \cdots = t_{k+11} = 1 + 1/k$, we find

$$\left(1 + \frac{1}{k}\right)^k \leq \left(\frac{11}{10}\right)^{11} = (1.1)^{11} < 2.854$$

5.26 By (5.64) with $p = 10$, we have $\alpha = 0.1$, and so

$$2A_0 = A_0 \left(1 + \frac{0.1}{k}\right)^{k \cdot n}$$

Dividing by A_0 and taking logarithms, we have

$$\log 2 = n \cdot k \cdot \log\left(1 + \frac{0.1}{k}\right)$$

or

$$n = \frac{\log 2}{k \log\left(1 + \frac{0.1}{k}\right)}$$

When interest is compounded annually, $k = 1$, and we find that $n = 7.2725$. This means that 8 years are needed to surpass twice the initial amount. Quarterly compounding of interest implies $k = 4$ and so $n = 7.0178$; that is, 7 years and 3 months are sufficient in this case. For monthly compounding we have $k = 12$, and thus $n = 6.960$. Hence 7 years are sufficient. For continuous compounding, we find from (5.66) that n must satisfy

$$2 = e^{(0.1) \cdot n}$$

or

$$n = 10 \cdot \ln\ 2 = 6.9315$$

Finally, to find the effective annual yield p corresponding to 10 percent compounded continuously, we have

$$1 + \frac{p}{100} = e^{0.1}$$

or

$$p = 100(e^{0.1} - 1) = 10.517$$

This means that 10 percent compounded continuously is effectively equivalent to about 10.5 percent.

5.27 By (5.72) with $\alpha = 0.1$, we have

$$\frac{1}{4} = \left(1 - \frac{0.1}{k}\right)^{k \cdot n}$$

leading to

$$n = \frac{\log 0.25}{k \log\left(1 - \frac{0.1}{k}\right)}$$

Setting successively $k = 1, 4, 12$, we find that $n = 13.1576$, $n = 13.6889$, and $n = 13.8050$, respectively. For continuous depreciation, we find from (5.73) that n must satisfy

$$\frac{1}{4} = e^{-(0.1) \cdot n}$$

or

$$n = 10 \cdot \ln\ 4 = 13.8630$$

Finally, to find the effective annual loss corresponding to 10 percent continuous depreciation, we have

$$1 - \frac{p}{100} = e^{-0.1}$$

or

$$p = 100(1 - e^{-0.1}) = 9.516$$

This means that 10 percent continuous depreciation is effectively equivalent to about 9.5 percent.

5.28 The sequence is clearly increasing since

$$c_n - c_{n-1} = \frac{1}{n^2} > 0$$

Moreover, $c_1 = 1$, and so $1 \leq c_n$ for all n. To show that the sequence is also bounded from above, we write

$$\begin{aligned} c_n &= \sum_{k=1}^{n} \frac{1}{k^2} = 1 + \sum_{k=2}^{n} \frac{1}{k^2} < 1 + \sum_{k=2}^{n} \frac{1}{(k-1)k} \\ &= 1 + \sum_{k=2}^{n} \left(\frac{1}{k-1} - \frac{1}{k}\right) \\ &= 1 + 1 - \frac{1}{2} + \frac{1}{2} - \frac{1}{3} + \cdots + \frac{1}{n-1} - \frac{1}{n} \\ &= 1 + 1 - \frac{1}{n} \leq 2 \end{aligned}$$

5.29 By (5.92) we can write

$$|e - A| = \left|e - \frac{163}{60}\right| = \left|e - \left(1 + \frac{1}{1!} + \frac{1}{2!} + \frac{1}{3!} + \frac{1}{4!} + \frac{1}{5!}\right)\right|$$

$$< \frac{\frac{1}{5}}{5!} = \frac{1}{600} = 0.001666$$

5.30 By (5.92), it is sufficient for k to satisfy

$$\frac{\frac{1}{k}}{k!} \leq \frac{1}{2} \cdot 10^{-q}$$

or

$$2 \cdot 10^{q} \leq k \cdot k!$$

Hence, (a) for $q = 4$, we have $k = 7$; (b) for $q = 8$, we have $k = 11$; and (c) for $q = 12$, we get $k = 15$.

Chapter 6

6.1 (a) $\alpha k^2 + \beta k + \gamma = O(k^2)$, $\alpha \neq 0$, because $(\alpha k^2 + \beta k + \gamma)/k^2 = \alpha + \beta/k + \gamma/k^2 \to \alpha$.

(b) $A\sqrt{k} + B\sqrt[3]{k} = O(\sqrt{k})$ because $(A\sqrt{k} + B\sqrt[3]{k})/\sqrt{k} = A + B/\sqrt[6]{k} \to A$.

(c) $D/(2k+9)^2 = O(1/k^2)$ because $\left[D/(2k+9)^2\right] / \left[1/k^2\right] = Dk^2/(4k^2 + 36k + 81) = D/(4 + 36/k + 81/k^2) \to D/4$.

(d) $k^2/(7 + \sqrt{k}) = O(k^{3/2})$ because $[k^2/(7 + \sqrt{k})]/k^{3/2} = k^2/(7k^{3/2} + k^2) = 1/(7/\sqrt{k} + 1) \to 1$.

6.2 (a) $\sin k = o(k)$, since $(\sin k)/k \to 0$ because we have $|\sin k| \leq 1$.

(b) $1/k! = o(2^{-k})$ because $(1/k!)/2^{-k} = 2^k/k! = [(2 \cdot 2 \cdot 2)/(1 \cdot 2 \cdot 3)][(2 \cdot 2 \cdot \cdots \cdot 2)/(4 \cdot 5 \cdot \cdots \cdot k)] = (8/6)(2/4)(2/5)\cdots(2/k) < (8/6) \cdot (1/2)^{k-3} = (64/6)/2^k < 11/2^k \to 0$.

(c) $2^{-k} = o(1/k^2)$ because, as we presently show, $2^{-k}/k^{-2} = k^2/2^k \to 0$. Indeed, replacing k by $(k+1)$, we have, for every $k \geq 3$,

$$\frac{(k+1)^2}{2^{k+1}} = \frac{k^2}{2^k}\frac{(k+1)^2}{2k^2} = \left(\frac{1}{2}\right)\left(\frac{k^2}{2^k}\right)\left(1 + \frac{2}{k} + \frac{1}{k^2}\right)$$

$$< \frac{1.8}{2}\frac{k^2}{2^k} = \frac{9}{10}\frac{k^2}{2^k}$$

Similarly, $(k+2)^2/2^{k+2} < (k^2/2^k) \cdot (9/10)^2$, and so on. Now, for $k = 3$ we have $k^2/2^k = 9/8$, and hence for all $n > 3$

we can conclude that $n^2/2^n < (9/8)\cdot(9/10)^{n-3} = (9/8)\cdot(10^3/9^3)\cdot(9/10)^n < 2\cdot(9/10)^n \to 0$.

6.3 (a) Here $a_1 = 0.2$ and $r = 0.1$, so that $0.2222\ldots = S = 0.2/(1-0.1) = 2/9$.

(b) In this case we have $3.17171717\ldots = 3 + 0.17171717\ldots = 3 + 0.17/(1-0.01) = 3 + 0.17/0.99 = 314/99$.

(c) Again, we have $0.9358585858\ldots = 0.93 + 0.0058585858\ldots = 93/100 + 0.0058/(1-0.01) = 93/100 + 58/9900 = 9265/9900$.

6.4 (a) Set $S_n = 2 - 1/(n+1)^2$, so that

$$\begin{aligned}
&S_1 + (S_2 - S_1) + (S_3 - S_2) + \cdots + (S_n - S_{n-1}) + \cdots \\
&\quad = \frac{7}{4} + \left(\frac{17}{9} - \frac{7}{4}\right) + \left(\frac{31}{16} - \frac{17}{9}\right) + \cdots \\
&\qquad + \left(\frac{-1}{(n+1)^2} - \frac{-1}{n^2}\right) + \cdots \\
&\quad = \frac{7}{4} + \frac{5}{36} + \frac{7}{144} + \cdots + \frac{2^{n+1}}{n^2(n+1)^2} + \cdots \\
&\quad = \frac{7}{4} + \sum_{n=2}^{\infty} \frac{2n+1}{n^2(n+1)^2} = 1 + \sum_{n=1}^{\infty} \frac{2n+1}{n^2(n+1)^2}
\end{aligned}$$

(b) Set $S_n = 2 - 1/\sqrt{n+1}$, so that

$$\begin{aligned}
&S_1 + (S_2 - S_1) + (S_3 - S_2) + \cdots + (S_n - S_{n-1}) + \cdots \\
&\quad = (2 - \frac{1}{\sqrt{2}}) + \left(\frac{-1}{\sqrt{3}} - \frac{-1}{\sqrt{2}}\right) + \left(\frac{-1}{2} - \frac{-1}{\sqrt{3}}\right) + \cdots \\
&\qquad + \left(\frac{-1}{\sqrt{n+1}} - \frac{-1}{\sqrt{n}}\right) + \cdots \\
&\quad = \frac{4-\sqrt{2}}{2} + \frac{3\sqrt{2} - 2\sqrt{3}}{6} + \frac{4\sqrt{3} - 3\sqrt{4}}{12} + \cdots \\
&\qquad + \frac{(n+1)\sqrt{n} - n\sqrt{n+1}}{n(n+1)} + \cdots \\
&\quad = 1 + \sum_{n=1}^{\infty} \frac{(n+1)\sqrt{n} - n\sqrt{n+1}}{n(n+1)} \\
&\quad = 1 + \sum_{n=1}^{\infty} \frac{\sqrt{1+1/n} - 1}{\sqrt{n+1}}
\end{aligned}$$

6.5 Generalizing the method employed in (6.18), we must look for a constant c such that

$$\frac{1}{n(n+1)(n+2)} = \frac{c}{n(n+1)} - \frac{c}{(n+1)(n+2)}$$

or

$$1 = (n+2)c - nc$$

from which we find that $c = \frac{1}{2}$. Thus,

$$S = \sum_{n=1}^{\infty} \frac{1}{n(n+1)(n+2)} = \frac{1}{2} \sum_{n=1}^{\infty} \left[\frac{1}{n(n+1)} - \frac{1}{(n+1)(n+2)} \right]$$
$$= \frac{1}{2} \left[\frac{1}{2} - \frac{1}{6} + \frac{1}{6} - \frac{1}{12} + - \cdots \right] = \frac{1}{4}$$

6.6 Here

$$R_k = \sum_{n=k+1}^{\infty} \frac{1}{n^3} = \frac{1}{(k+1)^3} + \frac{1}{(k+2)^3} + \cdots$$
$$< \sum_{n=k+1}^{\infty} \frac{1}{(n-2)(n-1)n}$$
$$= \sum_{n=k+1}^{\infty} \left[\frac{\frac{1}{2}}{(n-2)(n-1)} - \frac{\frac{1}{2}}{(n-1)n} \right]$$
$$= \frac{1}{2} \left[\frac{1}{(k-1)k} - \frac{1}{k(k+1)} + \frac{1}{k(k+1)} - \frac{1}{(k+1)(k+2)} + - \cdots \right] = \frac{1}{2k(k-1)}$$

where the technique employed in Exercise 6.5 has been used again. On the other hand, we have

$$R_k = \sum_{n=k+1}^{\infty} \frac{1}{n^3} > \sum_{n=k+1}^{\infty} \frac{1}{n(n+1)(n+2)}$$
$$= \sum_{n=k+1}^{\infty} \left[\frac{\frac{1}{2}}{n(n+1)} - \frac{\frac{1}{2}}{(n+1)(n+2)} \right]$$
$$= \frac{1}{2} \left[\frac{1}{(k+1)(k+2)} - \frac{1}{(k+2)(k+3)} + \frac{1}{(k+2)(k+3)} - \frac{1}{(k+3)(k+4)} + - \cdots \right] = \frac{1}{2(k+1)(k+2)}$$

Thus,

$$\frac{1}{2(k+1)(k+2)} < R_k < \frac{1}{2k(k-1)}$$

Using the average of these bounds, we can write

$$\begin{aligned} R_k &= \frac{1}{2}\left[\frac{1}{2(k-1)k} + \frac{1}{2(k+1)(k+2)}\right] + E \\ &= \frac{k^2+k+1}{2(k-1)k(k+1)(k+2)} + E \end{aligned}$$

where

$$\begin{aligned} |E| &< \frac{1}{2}\left[\frac{1}{2(k-1)k} - \frac{1}{2(k+1)(k+2)}\right] \\ &= \frac{2k+1}{2(k-1)k(k+1)(k+2)} \\ &= \frac{1}{(k-1)k(k+1)} \cdot \frac{k+\frac{1}{2}}{(k+2)} < \frac{1}{(k-1)k(k+1)} \\ &< \frac{1}{(k-1)^3} \end{aligned}$$

Collecting all the results, we obtain (6.26). Now, the original estimate of the remainder was $R_k < 1/[2k(k-1)] < 1/[2(k-1)^2]$, so six-figure accuracy requires $\frac{1}{2}/(k-1)^2 \le \frac{1}{2}\cdot 10^{-6}$, or $k \ge 1001$. On the other hand, after the addition of the correction term, we require $|E| < 1/(k-1)^3 \le \frac{1}{2}\cdot 10^{-6}$, or $(k-1)^3 \ge 2\cdot 10^6$, so that $k \ge 126$ will do.

6.7 Here,

$$\begin{aligned} R_k &= \sum_{n=k+1}^{\infty} \frac{1}{n^4} < \sum_{n=k+1}^{\infty} \frac{1}{(n-3)(n-2)(n-1)n} \\ &= \sum_{n=k+1}^{\infty} \left[\frac{\frac{1}{3}}{(n-3)(n-2)(n-1)} - \frac{\frac{1}{3}}{(n-2)(n-1)n}\right] \\ &= \frac{1}{3(k-2)(k-1)k} \end{aligned}$$

Similarly, for the lower bound,

$$R_k = \sum_{n=k+1}^{\infty} \frac{1}{n^4} > \sum_{n=k+1}^{\infty} \frac{1}{n(n+1)(n+2)(n+3)}$$

$$= \sum_{n=k+1}^{\infty} \left[\frac{\frac{1}{3}}{n(n+1)(n+2)} - \frac{\frac{1}{3}}{(n+1)(n+2)(n+3)} \right]$$

$$= \frac{1}{3(k+1)(k+2)(k+3)}$$

Thus,

$$\frac{1}{3(k+1)(k+2)(k+3)} < R_k < \frac{1}{3(k-2)(k-1)k}$$

and the averaging of the bounds yields

$$R_k = \frac{1}{2}\left[\frac{1}{3(k-2)(k-1)k} + \frac{1}{3(k+1)(k+2)(k+3)}\right] + E$$

$$= \frac{2k^3 + 3k^2 + 13k + 6}{6(k-2)(k-1)k(k+1)(k+2)(k+3)} + E$$

where

$$|E| < \frac{1}{2}\left[\frac{1}{3(k-2)(k-1)k} - \frac{1}{3(k+1)(k+2)(k+3)}\right]$$

$$= \frac{3k^2 + 3k + 2}{2(k-2)(k-1)k(k+1)(k+2)(k+3)}$$

$$< \frac{3(k+1)^2}{2(k-2)(k-1)k(k+1)(k+2)(k+3)}$$

$$< \frac{\frac{3}{2}}{(k-2)(k-1)k(k+1)} < \frac{\frac{3}{2}}{(k-2)^4}$$

Now, the original estimate of the remainder was $R_k < 1/[3(k-2)(k-1)k] < 1/[3(k-2)^3]$, so 12-figure accuracy requires $\frac{1}{3}/(k-2)^3 \leq \frac{1}{2} \cdot 10^{-12}$, or $(k-2)^3 \geq \frac{2}{3} \cdot 10^{12}$, so that $k \geq 8738$ will do. On the other hand, after the addition of the correction term we require $|E| < \frac{3}{2}/(k-2)^4 \leq \frac{1}{2} \cdot 10^{-12}$, or $(k-2)^4 \geq 3 \cdot 10^{12}$, so that $k \geq 1319$ will do.

6.9 To apply Kummer's acceleration we take the series $\sum 1/[n(n+1)]$, which is similar to the given one because

$$\frac{a_n}{b_n} = \frac{n^2}{n^4+1} \cdot \frac{n(n+1)}{1} = \frac{n^4+n^3}{n^4+1} = \frac{1+\dfrac{1}{n}}{1+\dfrac{1}{n^4}} \to 1$$

This similar series can be summed, and we have

$$\sum_{n=3}^{\infty} \frac{1}{n(n+1)} = \sum_{n=3}^{\infty} \left(\frac{1}{n} - \frac{1}{n+1}\right) = \frac{1}{3}$$

Following Kummer, we write

$$\begin{aligned}\sum_{n=3}^{\infty} \frac{n^2}{n^4+1} &= \sum_{n=3}^{\infty} \frac{1}{n(n+1)} + \sum_{n=3}^{\infty} \left[\frac{n^2}{n^4+1} - \frac{1}{n(n+1)}\right] \\ &= \frac{1}{3} + \sum_{n=3}^{\infty} \frac{n^3-1}{n(n+1)(n^4+1)}\end{aligned}$$

We now examine the remainder $\widetilde{R}_k$ of the new series. We have

$$\begin{aligned}\widetilde{R}_k &= \sum_{n=k+1}^{\infty} \frac{n^3-1}{n(n+1)(n^4+1)} \\ &< \sum_{n=k+1}^{\infty} \frac{n^3}{n(n+1)n^4} \\ &= \sum_{n=k+1}^{\infty} \frac{1}{n^2(n+1)}\end{aligned}$$

and so, from (6.29), $\widetilde{R}_k < 1(2k^2)$. Four-figure accuracy thus requires $\frac{1}{2}/k^2 \leq \frac{1}{2} \cdot 10^{-4}$, or $k \geq 100$. Thus, 98 terms of the new series must be summed because the series starts with $n = 3$. Had we not applied Kummer's acceleration but summed the given series as is, we would have

$$\begin{aligned}R_k = \sum_{n=k+1}^{\infty} \frac{n^2}{n^4+1} &< \sum_{n=k+1}^{\infty} \frac{1}{n^2} < \sum_{n=k+1}^{\infty} \frac{1}{(n-1)n} \\ &= \sum_{n=k+1}^{\infty} \left(\frac{1}{n-1} - \frac{1}{n}\right) = \frac{1}{k}\end{aligned}$$

Thus, requiring $1/k \leq \frac{1}{2} \cdot 10^{-4}$, we find $k \geq 20,000$, or 19,998 terms must be summed. This should be compared with 98 terms of the new series in which, however, each term is computationally more complicated than a typical term in the given series. If we want to apply Kummer's acceleration once again, we can take the series $\sum 1/[n(n+1)(n+2)]$, which is similar to the series $\sum[n^3-1]/[n(n+1)(n^4+1)]$ obtained through the first Kummer acceleration. This similarity follows from the fact that

$$\frac{n^3-1}{n(n+1)(n^4+1)} \cdot \frac{n(n+1)(n+2)}{1}$$
$$= \frac{n^4+2n^3-n-2}{n^4+1} = \frac{1+2/n-1/n^3-2/n^4}{1+1/n^4} \to 1$$

The series $\sum 1/[n(n+1)(n+2)]$ can be summed, and we have

$$\sum_{n=3}^{\infty} \frac{1}{n(n+1)(n+2)} = \sum_{n=3}^{\infty} \left[\frac{\frac{1}{2}}{n(n+1)} - \frac{\frac{1}{2}}{(n+1)(n+2)} \right] = \frac{1}{24}$$

Following Kummer again, we write

$$\frac{1}{3} + \sum_{n=3}^{\infty} \frac{n^3-1}{n(n+1)(n^4+1)}$$
$$= \frac{1}{3} + \frac{1}{24} + \sum_{n=3}^{\infty} \left[\frac{n^3-1}{n(n+1)(n^4+1)} - \frac{1}{n(n+1)(n+2)} \right]$$
$$= \frac{9}{24} + \sum_{n=3}^{\infty} \frac{2n^3-n-3}{n(n+1)(n+2)(n^4+1)}$$

The remainder $\tilde{\tilde{R}}_k$ of the last series will now be bounded. Indeed,

$$\tilde{\tilde{R}}_k = \sum_{n=k+1}^{\infty} \frac{2n^3-n-3}{n(n+1)(n+2)(n^4+1)}$$
$$< \sum_{n=k+1}^{\infty} \frac{2n^3}{n(n+1)(n+2)n^4}$$
$$< \sum_{n=k+1}^{\infty} \frac{2}{(n-1)n(n+1)(n+2)}$$
$$= 2 \sum_{n=k+1}^{\infty} \left[\frac{\frac{1}{3}}{(n-1)n(n+1)} - \frac{\frac{1}{3}}{n(n+1)(n+2)} \right]$$
$$= \frac{\frac{2}{3}}{k(k+1)(k+2)} < \frac{\frac{2}{3}}{k^3}$$

Four-figure accuracy thus requires $\frac{2}{3}/k^3 \le \frac{1}{2} \cdot 10^{-4}$, or $k^3 \ge \frac{4}{3} \cdot 10^4$, so that $k \ge 24$. Thus, only 22 terms of the final series must be summed.

6.10 We can write

$$\sum_{n=1}^{\infty} \frac{1}{n^3} = \sum_{n=1}^{\infty} \frac{1}{n(n+1)(n+2)} + \sum_{n=1}^{\infty} \left[\frac{1}{n^3} - \frac{1}{n(n+1)(n+2)}\right]$$

$$= \sum_{n=1}^{\infty} \left[\frac{\frac{1}{2}}{n(n+1)} - \frac{\frac{1}{2}}{(n+1)(n+2)}\right]$$

$$+ \sum_{n=1}^{\infty} \frac{3n+2}{n^3(n+1)(n+2)}$$

$$= \frac{1}{4} + \sum_{n=1}^{\infty} \frac{3n+2}{n^3(n+1)(n+2)}$$

To see the improvement, we examine the remainder $\widetilde{R}_k$ of the new series compared with the remainder R_k of the original series. Indeed,

$$\widetilde{R}_k = \sum_{n=k+1}^{\infty} \frac{3n+2}{n^3(n+1)(n+2)} < \sum_{n=k+1}^{\infty} \frac{3(n+2)}{n^3(n+1)(n+2)}$$

$$< 3 \sum_{n=k+1}^{\infty} \frac{1}{(n-2)(n-1)n(n+1)}$$

$$= 3 \sum_{n=k+1}^{\infty} \left[\frac{\frac{1}{3}}{(n-2)(n-1)n} - \frac{\frac{1}{3}}{(n-1)n(n+1)}\right]$$

$$= \frac{1}{(k-1)k(k+1)} < \frac{1}{(k-1)^3}$$

and

$$R_k = \sum_{n=k+1}^{\infty} \frac{1}{n^3} < \sum_{n=k+1}^{\infty} \frac{1}{(n-2)(n-1)n}$$

$$= \sum_{n=k+1}^{\infty} \left[\frac{\frac{1}{2}}{(n-2)(n-1)} - \frac{\frac{1}{2}}{(n-1)n}\right]$$

$$= \frac{1}{2(k-1)k} < \frac{1}{2(k-1)^2}$$

We find that six-figure accuracy requires 1001 terms for the original series but only 127 terms for the accelerated one.

If we want to apply Kummer's acceleration once again, we can write

$$\frac{1}{4}+\sum_{n=1}^{\infty}\frac{3n+2}{n^3(n+1)(n+2)}=\frac{1}{4}+\sum_{n=1}^{\infty}\frac{3}{n(n+1)(n+2)(n+3)}$$
$$+\sum_{n=1}^{\infty}\left[\frac{3n+2}{n^3(n+1)(n+2)}-\frac{3}{n(n+1)(n+2)(n+3)}\right]$$
$$=\frac{1}{4}+\sum_{n=1}^{\infty}\left[\frac{1}{n(n+1)(n+2)}-\frac{1}{(n+1)(n+2)(n+3)}\right]$$
$$+\sum_{n=1}^{\infty}\frac{11n+6}{n^3(n+1)(n+2)(n+3)}$$
$$=\frac{5}{12}+\sum_{n=1}^{\infty}\frac{11n+6}{n^3(n+1)(n+2)(n+3)}$$

The remainder $\tilde{\tilde{R}}_k$ of the last series can be bounded by

$$\tilde{\tilde{R}}_k=\sum_{n=k+1}^{\infty}\frac{11n+6}{n^3(n+1)(n+2)(n+3)}<\sum_{n=k+1}^{\infty}\frac{11}{n^3(n+1)(n+2)}$$
$$<\sum_{n=k+1}^{\infty}\frac{11}{(n-2)(n-1)n(n+1)(n+2)}$$
$$=11\sum_{n=k+1}^{\infty}\left[\frac{\frac{1}{4}}{(n-2)(n-1)n(n+1)}\right.$$
$$\left.-\frac{\frac{1}{4}}{(n-1)n(n+1)(n+2)}\right]$$
$$=\frac{\frac{11}{4}}{(k-1)k(k+1)(k+2)}<\frac{\frac{11}{4}}{(k-1)^4}$$

Thus, the twice-accelerated series necessitates a mere 50 terms to yield six-figure accuracy.

6.11 Here we have $(a_n/b_n) \to C \neq 0$, so that $[a_n/(Cb_n)] \to 1$ and $(1-Cb_n/a_n) \to 0$. Accordingly, we first note that $\sum Cb_n = CB$ and then modify (6.34) through

$$\begin{aligned}\sum_{n=1}^{\infty} a_n &= \sum_{n=1}^{\infty} Cb_n + \sum_{n=1}^{\infty} (a_n - Cb_n) \\ &= CB + \sum_{n=1}^{\infty} \left(1 - \frac{Cb_n}{a_n}\right) a_n = CB + \sum_{n=1}^{\infty} \widetilde{a}_n\end{aligned}$$

where $\widetilde{a}_n = (1 - Cb_n/a_n)a_n$ decays faster to zero than a_n, thus accelerating the convergence.

6.12 The assumed recursion relation takes the form

$$\frac{n+2}{2^{n+1}} = \alpha\frac{n+1}{2^n} + \beta\frac{n}{2^{n-1}}$$

Clearing the fractions leads to

$$(2\alpha + 4\beta - 1)n + (2\alpha - 2) = 0$$

and because this must hold for all $n = 0, 1, 2, \dots$, we obtain first $\alpha = 1$ and subsequently $\beta = -\frac{1}{4}$. We can therefore write

$$\frac{n+2}{2^{n+1}} = \frac{n+1}{2^n} - \frac{1}{4} \cdot \frac{n}{2^{n-1}}$$

and then sum both sides for $n = 1, 2, \dots$ and obtain

$$S - 1 - 1 = (S - 1) - \frac{1}{4}S$$

which is equivalent to (6.37). Solving for S, we find $S = 4$.

6.13 The assumed recursion relation takes the form

$$\frac{(n+2)^2+3}{5^{n+2}} = \alpha\frac{(n+1)^2+3}{5^{n+1}} + \beta\frac{n^2+3}{5^n} + \gamma\frac{(n-1)^2+3}{5^{n-1}}$$

where the left-hand side is clearly a_{n+2}. Clearing the fractions leads to

$$\begin{aligned}(5\alpha + 25\beta + 125\gamma - 1)n^2 + (10\alpha - 250\gamma - 4)n \\ + (20\alpha + 75\beta + 500\gamma - 7) = 0\end{aligned}$$

The coefficient of each power of n must vanish because this equation must hold for all $n = 0, 1, 2, \dots$, and thus the middle term yields $\alpha = 25\gamma + \frac{2}{5}$. Substituting this expression for α in the remaining two equations, we obtain two linear equations for β and γ, whose solutions are $\beta = -\frac{3}{25}$ and $\gamma = \frac{1}{125}$, which in turn yields $\alpha = \frac{3}{5}$. Now we can write

$$\frac{(n+2)^2+3}{5^{n+2}} = \frac{3}{5}\cdot\frac{(n+1)^2+3}{5^{n+1}} - \frac{3}{25}\cdot\frac{n^2+3}{5^n} + \frac{1}{125}\cdot\frac{(n-1)^2+3}{5^{n-1}}$$

and then sum both sides for $n = 1, 2, \dots$ and obtain

$$S - 3 - \frac{4}{5} - \frac{7}{25} = \frac{3}{5}\left(S - 3 - \frac{4}{5}\right) - \frac{3}{25}(S-3) + \frac{1}{125}S$$

Solving for S, we find $S = \frac{135}{32}$.

6.14 Here we have $a_n = (-1)^n/(n+1)$, so that (6.42) takes the form

$$\begin{aligned} a_{n+1} - \alpha a_n - \beta a_{n-1} \\ &= \frac{(-1)^{n+1}}{n+2} - \alpha\frac{(-1)^n}{n+1} - \beta\frac{(-1)^{n-1}}{n} \\ &= (-1)^n\frac{(\beta-\alpha-1)n^2 + (3\beta-2\alpha-1)n + 2\beta}{n(n+1)(n+2)} \end{aligned}$$

Choosing α and β so that the coefficients of n^2 and n vanish, we obtain $\alpha = -2$ and $\beta = -1$ and are left with

$$a_{n+1} + 2a_n + a_{n-1} = \frac{2(-1)^{n+1}}{n(n+1)(n+2)}$$

This approximate recursion relation satisfies

$$|a_{n+1} + 2a_n + a_{n-1}| = \frac{2}{n(n+1)(n+2)} < \frac{2}{n^3}$$

or

$$|a_{n+1} + 2a_n + a_{n-1}| = O\left(\frac{1}{n^3}\right)$$

6.15 This time we start with

$$\begin{aligned} a_{n+2} - \alpha a_{n+1} - \beta a_n - \gamma a_{n-1} \\ &= (-1)^n\left[(\alpha-\beta+\gamma+1)n^3 + (4\alpha-5\beta+6\gamma+3)n^2\right. \\ &\quad \left. + (3\alpha-6\beta+11\gamma+2)n + 6\gamma\right] \Big/ n(n+1)(n+2)(n+3) \end{aligned}$$

Choosing α, β, and γ so that the coefficients of n^3, n^2, and n vanish, we get $\alpha = -3$, $\beta = -3$, and $\gamma = -1$, and are left with

$$a_{n+2} + 3a_{n+1} + 3a_n + a_{n-1} = \frac{6(-1)^{n+1}}{n(n+1)(n+2)(n+3)}$$

This approximate recursion relation satisfies

$$|a_{n+2} + 3a_{n+1} + 3a_n + a_{n-1}| = \frac{6}{n(n+1)(n+2)(n+3)} < \frac{6}{n^4}$$

or

$$|a_{n+2} + 3a_{n+1} + 3a_n + a_{n-1}| = O\left(\frac{1}{n^4}\right)$$

For the two-term approximate recursion relation we have

$$a_{n+1} - \alpha a_n = (-1)^{n+1} \frac{(\alpha+1)n + (2\alpha+1)}{(n+1)(n+2)}$$

Choosing $\alpha = -1$, we are left with

$$a_{n+1} + a_n = \frac{(-1)^n}{(n+1)(n+2)}$$

which satisfies

$$|a_{n+1} + a_n| = \frac{1}{(n+1)(n+2)} < \frac{1}{n^2}$$

or

$$|a_{n+1} + a_n| = O\left(\frac{1}{n^2}\right)$$

6.16 In Exercise 6.14 we found the approximate three-term recursion relation

$$a_{n+1} + 2a_n + a_{n-1} = \frac{2(-1)^{n+1}}{n(n+1)(n+2)}$$

Summing both sides for $n = 1, 2, \ldots$, we obtain

$$(S - a_0 - a_1) + 2(S - a_0) + S = 2 \sum_{n=1}^{\infty} \frac{(-1)^{n+1}}{n(n+1)(n+2)}$$

Because $a_0 = 1$ and $a_1 = -\frac{1}{2}$, we find that

$$S = \frac{5}{8} + \frac{1}{2} \sum_{n=1}^{\infty} \frac{(-1)^{n+1}}{n(n+1)(n+2)}$$

Now the remainder of the original series, after k terms, satisfies

$$|R_k| = \left| \sum_{n=k+1}^{\infty} \frac{(-1)^n}{n+1} \right| \leq \frac{1}{k+2} < \frac{1}{k}$$

because its terms decrease in absolute value and alternate in sign, and similarly the remainder of the accelerated series satisfies

$$|\widetilde{R}_k| = \left|\frac{1}{2}\sum_{n=k+1}^{\infty}\frac{(-1)^{n+1}}{n(n+1)(n+2)}\right|$$
$$\leq \frac{\frac{1}{2}}{(k+1)(k+2)(k+3)} < \frac{1}{2k^3}$$

Next, to assure q decimal figures in the original series, we need $k \geq 2\cdot 10^q$, that is, 20,000 terms for $q = 4$ and 200,000,000 terms for $q = 8$, whereas for the accelerated series we need $k \geq 10^{q/3}$, that is, 22 terms for $q = 4$ and 465 terms for $q = 8$. The effectiveness of the acceleration for this case is impressive indeed.

6.17 In Exercise 6.15 we found the approximate four-term recursion relation

$$a_{n+2} + 3a_{n+1} + 3a_n + a_{n-1} = \frac{6(-1)^{n+1}}{n(n+1)(n+2)(n+3)}$$

Summing both sides for $n = 1, 2, \ldots$ and noting that $a_0 = 1$, $a_1 = -\frac{1}{2}$, and $a_2 = \frac{1}{3}$, we reach

$$S = \frac{2}{3} + \frac{3}{4}\sum_{n=1}^{\infty}\frac{(-1)^{n+1}}{n(n+1)(n+2)(n+3)}$$

This time, the remainder after k terms satisfies

$$\left|\frac{3}{4}\sum_{n=k+1}^{\infty}\frac{(-1)^{n+1}}{n(n+1)(n+2)(n+3)}\right|$$
$$\leq \frac{\frac{3}{4}}{(k+1)(k+2)(k+3)(k+4)} < \frac{\frac{3}{4}}{k^4}$$

and thus to assure q decimal figures we need $k \geq \sqrt[4]{1.5}\cdot 10^{q/4}$, or 12 terms for $q = 4$ and 111 terms for $q = 8$. Comparing these results with the results of Exercise 6.16, we see that for $q = 4$, the number of terms needed was reduced by a factor of about 2 whereas for $q = 8$, the reduction factor is about 4.

6.20 To obtain an approximate three-term recursion relation, we write

$$a_{n+1} - \alpha a_n - \beta a_{n-1} = r^{n-1}\frac{(r^2-\alpha r-\beta)n^2+(r^2-2\alpha r-3\beta)n-2\beta}{n(n+1)(n+2)}$$

and require that $r^2 - \alpha r - \beta = 0$ and $r^2 - 2\alpha r - 3\beta = 0$. Thus, $\alpha = 2r$, $\beta = -r^2$, and we have

$$a_{n+1} - 2ra_n + r^2 a_{n-1} = \frac{2r^{n+1}}{n(n+1)(n+2)}$$

Summing for $n = 1, 2\ldots$ and noting also that $a_0 = 1$ and $a_1 = r/2$, we reach

$$S = \sum_{n=0}^{\infty}\frac{r^n}{n+1} = \frac{1-\left(\frac{3}{2}\right)r}{(1-r)^2} + \frac{2}{(1-r)^2}\sum_{n=1}^{\infty}\frac{r^{n+1}}{n(n+1)(n+2)}$$

which is the analog of (6.60) in the text. Because we are interested here only in $-1 \le r < 0$, the remainder of the original series can be estimated by

$$|R_k| = \left|\sum_{n=k+1}^{\infty}\frac{r^n}{n+1}\right| \le \frac{|r|^{k+1}}{k+2} < \frac{|r|^{k+1}}{k}$$

whereas for the accelerated series we have

$$|\widetilde{R}_k| = \left|\frac{2}{(1-r)^2}\sum_{n=k+1}^{\infty}\frac{r^{n+1}}{n(n+1)(n+2)}\right| \le \frac{2|r|^{k+2}}{(1-r)^2(k+1)(k+2)(k+3)} < \frac{2|r|^{k+2}}{(1-r)^2k^3}$$

For $r = -1$, we have $|R_k| < 1/k \le \frac{1}{2}\cdot 10^{-4}$ to assure four decimal figures, or $k \ge 20{,}000$ terms are needed, whereas $|\widetilde{R}_k| < 2/(4k^3) \le \frac{1}{2}\cdot 10^{-4}$, or $k \ge 22$. Next, for $r = -0.9$, we have $|R_k| < (0.9)^{k+1}/k \le \frac{1}{2}\cdot 10^{-4}$, which holds for $k \ge 55$, whereas $|\widetilde{R}_k| < 2\cdot(0.9)^{k+2}/\left[(1.9)^2\cdot k^3\right] \le \frac{1}{2}\cdot 10^{-4}$, which necessitates $k \ge 14$.

6.21 This time we have

$$a_{n+2} - \alpha a_{n+1} - \beta a_n - \gamma a_{n-1} = r^{n-1}\frac{An^3+Bn^2+Cn-6\gamma}{n(n+1)(n+2)(n+3)}$$

where

$$A = r^3 - \alpha r^2 - \beta r - \gamma$$
$$B = 3r^3 - 4\alpha r^2 - 5\beta r - 6\gamma$$
$$C = 2r^3 - 3\alpha r^2 - 6\beta r - 11\gamma$$

Next we choose α, β, and γ so that $A = B = C = 0$. This yields $\alpha = 3r$, $\beta = -3r^2$, and $\gamma = r^3$, so that we have

$$a_{n+2} - 3ra_{n+1} + 3r^2 a_n - r^3 a_{n-1} = \frac{-6r^{n+2}}{n(n+1)(n+2)(n+3)}$$

Summing for $n = 1, 2, \ldots$ and noting that $a_0 = 1$, $a_1 = r/2$, and $a_2 = r^2/3$, we reach

$$S = \frac{1 - \frac{5}{2}r + \frac{11}{6}r^2}{(1-r)^3} - \frac{6}{(1-r)^3} \sum_{n=1}^{\infty} \frac{r^{n+2}}{n(n+1)(n+2)(n+3)}$$

Because we are interested here only in $-1 \le r < 0$, the remainder can be estimated by

$$|\tilde{\tilde{R}}_k| \le \left| \frac{6r^{k+3}}{(1-r)^3(k+1)(k+2)(k+3)(k+4)} \right| < \frac{6|r|^{k+3}}{(1-r)^3 \cdot k^4}$$

For $r = -1$, we have $|\tilde{\tilde{R}}_k| < 6/(8k^4) \le \frac{1}{2} \cdot 10^{-4}$, to assure four decimal figures, that is, $k \ge 12$ terms are needed. Next, for $r = -0.9$, we have $|\tilde{\tilde{R}}_k| < 6 \cdot (0.9)^{k+3} / \left[(1.9)^3 \cdot k^4\right] \le \frac{1}{2} \cdot 10^{-4}$, which necessitates $k \ge 9$.

6.23 The remainder of the series is

$$R_k = \sum_{n=k+1}^{\infty} \frac{r^n}{(n+1)^p}$$

where p is a positive integer, and this time we are interested in $0 < r < 1$. To begin with, we can estimate the remainder by

$$0 < R_k = \sum_{n=k+1}^{\infty} \frac{r^n}{(n+1)^p}$$
$$\le \sum_{n=k+1}^{\infty} \frac{r^n}{(n-p+2)(n-p+3)\cdots n(n+1)}$$

Next, we want to multiply the numerator by the appropriate factor F, which will turn the series on the right into a telescoping series. From (6.71) we find that in our case F is given by

$$F = \frac{(n+1) - (n-p+2)r}{p-1} = 1 + \frac{n-p+2}{p-1}(1-r) > 1$$

because the largest factor of our denominator is $(n+1)$, yielding $i = 1$, whereas the smallest factor is $(n-p+2)$, so that $j = p-2$. Multiplying the series on the right by this factor F, we reach

$$\begin{aligned} 0 < R_k &\leq \sum_{n=k+1}^{\infty} \frac{\big[(n+1) - (n-p+2)r\big] r^n}{(n-p+2)(n-p+3)\cdots n(n+1)} \\ &= \sum_{n=k+1}^{\infty} \left[\frac{r^n}{(n-p+2)(n-p+3)\cdots n} \right. \\ &\qquad \left. - \frac{r^{n+1}}{(n-p+3)\cdots n(n+1)}\right] \end{aligned}$$

We see that this is a telescoping series, whose easily computed sum $r^{k+1}/\,[(k-p+3)(k-p+4)\cdots(k+1)]$ bounds R_k. Thus,

$$0 < R_k \leq \frac{r^{k+1}}{(k-p+3)(k-p+4)\cdots(k+1)}$$

Observe that for $p = 2$ the denominator consists of the single factor $(k+1)$, and the result reduces to (6.67). For $p = 5$, say, we obtain

$$0 < R_k \leq \frac{r^{k+1}}{(k-2)(k-1)k(k+1)} < \frac{r^{k+1}}{(k-2)^4}$$

Chapter 7

7.1 We have already established in the text that $\sin 15° = \cos 75° = \sqrt{2-\sqrt{3}}/2$ and that $\cos 15° = \sin 75° = \sqrt{2+\sqrt{3}}/2$. Accordingly,

$$= \sqrt{(1-\cos 15°)/2} = \frac{\sqrt{2-\sqrt{2+\sqrt{3}}}}{2} = \cos 82.5°$$

$$\cos 7.5^\circ = \sqrt{(1+\cos 15^\circ)/2} = \frac{\sqrt{2+\sqrt{2+\sqrt{3}}}}{2} = \sin 82.5^\circ$$

$$\cos 3.75^\circ = \sqrt{(1+\cos 7.5^\circ)/2} = \frac{\sqrt{2+\sqrt{2+\sqrt{2+\sqrt{3}}}}}{2}$$

$$= \sin 86.25^\circ$$

Also,

$$\sin 22.5^\circ = \sqrt{(1-\cos 45^\circ)/2} = \sqrt{\frac{1-\dfrac{\sqrt{2}}{2}}{2}}$$

$$= \frac{\sqrt{2-\sqrt{2}}}{2}$$

7.2 Assume two distinct parabolas $P_2(x)$ and $Q_2(x)$ that pass through the three distinct given points (x_0, y_0), (x_1, y_1), and (x_2, y_2); that is, we have $P_2(x_j) = Q_2(x_j) = y_j$, for $j = 0, 1, 2$. The difference $P_2(x) - Q_2(x)$ is also a polynomial of (at most) second degree, obviously vanishing for the three distinct values x_0, x_1, and x_2 of x. This, of course, is impossible unless the difference $P_2(x) - Q_2(x)$ vanishes identically, that is, if $P_2(x)$ is identically equal to $Q_2(x)$.

7.3 First note that this exercise is a generalization of Exercise 7.2. We now recall the well-known result that an nth-degree polynomial has at most n real roots. As in Exercise 7.2, we assume that two distinct polynomials $P_n(x)$ and $Q_n(x)$, of nth degree at most, pass through the given $(n+1)$ distinct points (x_0, y_0), $(x_1, y_1), \ldots, (x_n, y_n)$, that is, $P_n(x_j) = Q_n(x_j) = y_j$ for $j = 0, 1, \ldots, n$. The difference $P_n(x) - Q_n(x)$ is also a polynomial of (at most) nth degree, obviously vanishing for the $(n+1)$ distinct values x_0, $x_1, \ldots, x_n$ of x. By the preceding result, this is impossible unless the difference $P_n(x) - Q_n(x)$ vanishes identically, that is, if $P_n(x)$ is identically equal to $Q_n(x)$.

7.4 The third-degree Lagrange interpolation polynomia that passes through the four distinct points (x_0, y_0), (x_1, y_1), (x_2, y_2), and

(x_3, y_3), has the form

$$P_3(x) = \frac{(x-x_1)(x-x_2)(x-x_3)}{(x_0-x_1)(x_0-x_2)(x_0-x_3)}y_0 + \frac{(x-x_0)(x-x_2)(x-x_3)}{(x_1-x_0)(x_1-x_2)(x_1-x_3)}y_1 + \frac{(x-x_0)(x-x_1)(x-x_3)}{(x_2-x_0)(x_2-x_1)(x_2-x_3)}y_2 + \frac{(x-x_0)(x-x_1)(x-x_2)}{(x_3-x_0)(x_3-x_1)(x_3-x_2)}y_3$$

In analogy with (7.4) in the text, we denote the denominators by D_j, so we will have $D_0 = (x_0-x_1)(x_0-x_2)(x_0-x_3)$, $D_1 = (x_1-x_0)(x_1-x_2)(x_1-x_3)$, $D_2 = (x_2-x_0)(x_2-x_1)(x_2-x_3)$, and $D_3 = (x_3-x_0)(x_3-x_1)(x_3-x_2)$. Now we rewrite $P_3(x)$ in the form $P_3(x) = a_3x^3 + a_2x^2 + a_1x + a_0$, carry out the indicated multiplications in the four numerators, and collect coefficients of like powers. This leads to

$$a_3 = \frac{1}{D_0} + \frac{1}{D_1} + \frac{1}{D_2} + \frac{1}{D_3}$$

$$a_2 = -\frac{x_1+x_2+x_3}{D_0} - \frac{x_0+x_2+x_3}{D_1} - \frac{x_0+x_1+x_3}{D_2} - \frac{x_0+x_1+x_2}{D_3}$$

$$a_1 = \frac{x_1x_2+x_2x_3+x_3x_1}{D_0} + \frac{x_0x_2+x_2x_3+x_3x_0}{D_1} + \frac{x_0x_1+x_1x_3+x_3x_0}{D_2} + \frac{x_0x_1+x_1x_2+x_2x_0}{D_3}$$

$$a_0 = -\frac{x_1x_2x_3}{D_0} - \frac{x_0x_2x_3}{D_1} - \frac{x_0x_1x_3}{D_2} - \frac{x_0x_1x_2}{D_3}$$

7.8 (a) We have $y = m(x-x_0) + y_0$, where the slope is given by $m = (y_1-y_0)/(x_1-x_0) = -\frac{1}{4}$, so that $y = -\frac{1}{4}(x-1)+1 = -\frac{1}{4}x + \frac{5}{4}$.

(b) The line must be of the form $y = -\frac{1}{4}x + c$, and we require that it should coincide with $y = 1/x$ at exactly one point. This means that the equation $-\frac{1}{4}x + c = 1/x$ has only one (double) solution; that is, the quadratic equation $x^2 - 4cx + 4 = 0$ has only one (double) root. To ensure this, the discriminant $(16c^2 - 16)$ must vanish, leading to $c = \pm 1$ with corresponding values $x = \pm 2$. Because we are restricted to $1 \leq x \leq 4$, we must choose $c = 1$, and the required line is $y = -\frac{1}{4}x + 1$.

(c) As we saw in part (b), $c = 1$ and the relevant quadratic equation is $x^2 - 4x + 4 = 0$. Its solution is $x = 2$, for which $y = 1/x = -\frac{1}{4}x + 1 = \frac{1}{2}$. Hence the line is tangent to $1/x$ at $x = 2$.

(d) The required midway line is $P_1(x) = -\frac{1}{4}x + d$, where $d = (\frac{5}{4} + 1)/2 = \frac{9}{8}$ is the average of the y-intercepts of the lines in (a) and (b). Thus, $P_1(x) = -\frac{1}{4}x + \frac{9}{8}$.

(e) For $R(x) = 1/x - P_1(x) = 1/x + x/4 - 9/8$, we have $R(1) = -R(2) = R(4) = \frac{1}{8}$, thus verifying the required equal ripple property.

(g) We have $1/x = -\frac{1}{4}x + \frac{9}{8}$, or $2x^2 - 9x + 8 = 0$, with solutions $x_0 = (9 - \sqrt{17})/4 \approx 1.219$ and $x_1 = (9 + \sqrt{17})/4 \approx 3.281$. These are the two interpolation points at which the function $1/x$ and the interpolating line $P_1(x)$ coincide.

(h) In this case the tangent line to the curve $y = \sqrt{x}$ is $y = x/3 + c$, with c determined so that the equation $\sqrt{x} = x/3 + c$ has only one (double) solution. With $t = \sqrt{x}$, this means that the equation $t^2 - 3t + 3c = 0$ has only one (double) root, that is, $c = \frac{3}{4}$. For this value of c, $t = \frac{3}{2}$, and hence $x = \frac{9}{4}$, at which the line $y = x/3 + 3/4$ is tangent to $y = \sqrt{x}$. Now, for $R(x) = \sqrt{x} - (x/3 + 17/24)$ we find $R(1) = -R(9/4) = R(4) = -1/24$, verifying the equal ripple property in this case. In this connection see Fig. 2.2 and the related results in Section 2.5.

7.9 We have $P_2(x) = P_1(x) + \gamma(x - x_0)(x - x_1)$, where $P_2(x_0) = P_1(x_0) = y_0$ and $P_2(x_1) = P_1(x_1) = y_1$. We require $P_2(x_2) = y_2$, or $y_2 = P_1(x_2) + \gamma(x_2 - x_1)(x_2 - x_0)$. Substituting for $P_1(x_2)$ from (7.9), we reach

$$\gamma(x_2 - x_0)(x_2 - x_1) = (y_2 - y_0) - \frac{y_1 - y_0}{x_1 - x_0}(x_2 - x_0)$$

that is,

$$\gamma = \frac{y_2 - y_0}{(x_2 - x_1)(x_2 - x_0)} - \frac{y_1 - y_0}{(x_2 - x_1)(x_1 - x_0)}$$
$$= \frac{\dfrac{y_2 - y_0}{x_2 - x_0} - \dfrac{y_1 - y_0}{x_1 - x_0}}{x_2 - x_1}$$

which is, of course, a divided difference of divided differences. Clearly, because this analysis made no use of the relative positions of the (distinct) interpolation points, we are at liberty to interchange, say, (x_0, y_0) and (x_1, y_1). This yields, immediately,

$$\gamma = \frac{\dfrac{y_2 - y_1}{x_2 - x_1} - \dfrac{y_1 - y_0}{x_1 - x_0}}{x_2 - x_0}$$

As a check, we can evaluate the difference of the two expressions obtained for γ and find that it vanishes identically.

7.10 It is sufficient to show that the coefficients of x^2 and x, as well as the constant term, in the parabolas represented by (7.18) and (7.19) are equal, respectively. The coefficient of x^2 in Newton's form (7.18), after some rearrangement, is

$$A = \frac{y_0(x_2 - x_1) - y_1(x_2 - x_0) + y_2(x_1 - x_0)}{(x_2 - x_1)(x_1 - x_0)(x_2 - x_0)}$$
$$= \frac{y_0}{(x_1 - x_0)(x_2 - x_0)} - \frac{y_1}{(x_2 - x_1)(x_1 - x_0)}$$
$$+ \frac{y_2}{(x_2 - x_1)(x_2 - x_0)}$$

On the other hand, the coefficient of x^2 in Lagrange's form (7.19) is

$$\widetilde{A} = \frac{y_0}{(x_0 - x_1)(x_0 - x_2)} + \frac{y_1}{(x_1 - x_0)(x_1 - x_2)}$$
$$+ \frac{y_2}{(x_2 - x_0)(x_2 - x_1)}$$

which coincides with the last expression, that is, $A = \widetilde{A}$. Next, the coefficient of x in (7.18) is

$$\begin{aligned}
B &= \frac{y_1 - y_0}{x_1 - x_0} - (x_0 + x_1)A \\
&= \frac{y_1 - y_0}{x_1 - x_0} - \frac{y_0(x_0 + x_1)}{(x_0 - x_1)(x_0 - x_2)} - \frac{y_1(x_0 + x_1)}{(x_1 - x_0)(x_1 - x_2)} \\
&\quad - \frac{y_2(x_0 + x_1)}{(x_2 - x_0)(x_2 - x_1)} \\
&= - y_0 \left[\frac{x_0 + x_1}{(x_0 - x_1)(x_0 - x_2)} - \frac{1}{x_0 - x_1}\right] \\
&\quad - y_1 \left[\frac{x_0 + x_1}{(x_1 - x_0)(x_1 - x_2)} - \frac{1}{x_1 - x_0}\right] \\
&\quad - y_2 \frac{x_0 + x_1}{(x_2 - x_0)(x_2 - x_1)}
\end{aligned}$$

On the other hand, the coefficient of x in (7.19) is

$$\begin{aligned}
\widetilde{B} = &- \frac{y_0(x_1 + x_2)}{(x_0 - x_1)(x_0 - x_2)} - \frac{y_1(x_0 + x_2)}{(x_1 - x_0)(x_1 - x_2)} \\
&- \frac{y_2(x_0 + x_1)}{(x_2 - x_0)(x_2 - x_1)}
\end{aligned}$$

A little reflection shows that $B = \widetilde{B}$. Finally, the constant term in (7.18) is

$$\begin{aligned}
C &= y_0 - x_0 \frac{y_1 - y_0}{x_1 - x_0} + x_0 x_1 \cdot A \\
&= y_0 \left[1 + \frac{x_0}{x_1 - x_0} + \frac{x_0 x_1}{(x_0 - x_1)(x_0 - x_2)}\right] \\
&\quad + y_1 \left[\frac{-x_0}{x_1 - x_0} + \frac{x_0 x_1}{(x_1 - x_0)(x_1 - x_2)}\right] \\
&\quad + y_2 \frac{x_0 x_1}{(x_2 - x_0)(x_2 - x_1)}
\end{aligned}$$

On the other hand, the constant term in (7.19) is

$$\begin{aligned}
\widetilde{C} = &\frac{y_0 x_1 x_2}{(x_0 - x_1)(x_0 - x_2)} + \frac{y_1 x_0 x_2}{(x_1 - x_0)(x_1 - x_2)} \\
&+ \frac{y_2 x_0 x_1}{(x_2 - x_0)(x_2 - x_1)}
\end{aligned}$$

If we combine the coefficients of y_0 and y_1, respectively, in the expression for C, we see that $C = \widetilde{C}$. This shows that both the the Lagrangian and Newton's representations are given by $P_2(x) = Ax^2 + Bx + C$.

7.11 We start by constructing $P_4(x)$, which we want to pass through points (x_0, y_0), ... , (x_3, y_3), and (x_4, y_4). Thus, we search for λ, which is to satisfy

$$P_4(x) = P_3(x) + \lambda \cdot (x - x_0)(x - x_1)(x - x_2)(x - x_3)$$

as well as $P_4(x_4) = y_4$. This leads to

$$\begin{aligned}\lambda &= \frac{y_4 - P_3(x_4)}{(x_4 - x_0)(x_4 - x_1)(x_4 - x_2)(x_4 - x_3)} \\ &= \lambda(f; x_0, x_1, x_2, x_3, x_4)\end{aligned}$$

and hence

$$\begin{aligned}P_4(x_4) - P_3(x_4) &= f(x_4) - P_3(x_4) \\ &= \lambda(f; x_0, x_1, x_2, x_3, x_4) \\ &\cdot (x_4 - x_0)(x_4 - x_1)(x_4 - x_2)(x_4 - x_3)\end{aligned}$$

Because x_4 could have been any point in $[a, b]$, we can rewrite the preceding in the form

$$\begin{aligned}R_3(x) &= f(x) - P_3(x) \\ &= \lambda(f; x_0, x_1, x_2, x_3, x) \\ &\qquad \cdot (x - x_0)(x - x_1)(x - x_2)(x - x_3),\end{aligned}$$

with

$$\lambda = \frac{y - P_3(x)}{(x - x_0)(x - x_1)(x - x_2)(x - x_3)}$$

Now we substitute for $P_3(x)$ from (7.23), in which $P_2(x)$ is given by (7.18) whereas δ is given by (7.22). This substitution, admittedly voluminous, yields

$$\lambda = \frac{\delta(f; x_0, x_1, x_2, x) - \delta(f; x_0, x_1, x_2, x_3)}{x - x_3}$$

Because δ is a divided difference of third order, we see that λ is a divided difference of fourth order.

7.15 Expanding (7.42), we have

$$x^3 - (2\alpha + \beta)x^2 + (\alpha^2 + 2\alpha\beta)x - \alpha^2\beta = 0$$

and comparing with (7.41), we find that $2\alpha+\beta=-p$, $\alpha^2+2\alpha\beta=q$, and $k=r+\alpha^2\beta$. The first two of these relations now imply that $3\alpha^2+2p\alpha+q=0$, or $\alpha=(-p\pm\sqrt{p^2-3q})/3$, and in turn $\beta=-p-2\alpha=(-p\mp2\sqrt{p^2-3q})/3$. Consequently, from the third relation we obtain

$$k=r+\alpha^2\beta=\frac{27r+2p^3-9pq}{27}\pm\frac{2(p^2-3q)^{3/2}}{27}$$

verifying (7.43).

7.16 By (7.40) we have

$$\begin{aligned}p^2-3q&=(x_0+x_1+x_2)^2-3(x_0x_1+x_1x_2+x_2x_0)\\&=x_0^2+x_1^2+x_2^2-x_0x_1-x_1x_2-x_2x_0\\&=\frac{x_0^2+x_1^2-2x_0x_1}{2}+\frac{x_1^2+x_2^2-2x_1x_2}{2}\\&\quad+\frac{x_2^2+x_0^2-2x_0x_2}{2}\\&=\frac{(x_0-x_1)^2+(x_1-x_2)^2+(x_2-x_0)^2}{2}>0\end{aligned}$$

since x_0, x_1, and x_2 are distinct.

7.21 Substituting $x_{\text{old}}=a$, we obtain $x_{\text{new}}=-1$. For $x_{\text{old}}=(b+a)/2$, we have $x_{\text{new}}=0$. Finally, $x_{\text{old}}=b$ yields $x_{\text{new}}=1$. Solving for x_{old} leads to $x_{\text{old}}=(b+a)/2+[(b-a)/2]x_{\text{new}}$, which carries $[-1,1]$ back to $[a,b]$. If we rewrite the original transformation in the form

$$x_{\text{new}}=\frac{2}{b-a}\left(x_{\text{old}}-\frac{b+a}{2}\right)$$

we see that the subtraction of $(b+a)/2$ from x_{old} is a displacement of the interval $[a,b]$, which in particular carries the interval's midpoint into the origin. The multiplication by the factor $2/(b-a)$ has the effect of stretching (or contracting) the displaced interval to make it coincide with $[-1,1]$.

7.22 In the transformation $x_{\text{old}}=(b+a)/2+[(b-a)/2]x_{\text{new}}$, that is, $x_{\text{old}}=[(b-a)/2][x_{\text{new}}+(b+a)/(b-a)]$, we substitute $x_{\text{new}}=-\sqrt{3}/2$, 0, and $\sqrt{3}/2$, successively. This gives (7.54). Now, the points $x_0=-\sqrt{3}/2$, $x_1=0$, and $x_2=\sqrt{3}/2$ were chosen so that the ripples of $v=(x-x_0)(x-x_1)(x-x_2)$ in the interval $[-1,1]$ will be equal, indeed equal to $\frac{1}{4}$. The displacement in the transformation has obviously no effect on the size of the ripples. On the other hand, the stretching (or contracting) factor $(b-a)/2$ is applied to each of the factors $(x-x_0), (x-x_1)$, and $(x-x_2)$ of v, and hence its overall effect is to multiply the original equal ripple size $\frac{1}{4}$ by $[(b-a)/2]^3$, as in (7.55).

7.29 To determine the values of x_0 and x_1 that will yield an equal rippled $E_2(x)$, we must require $u(a) = -u_{\min} = u(b)$, where u is given by $u = x^2 - (x_0 + x_1)x + x_0x_1$ in (7.35). Now, $u(a) = u(b)$, that is, $a^2 - (x_0+x_1)a + x_0x_1 = b^2 - (x_0+x_1)b + x_0x_1$, and so $x_0 + x_1 = a + b$. Moreover, from (7.37) we have $u_{\min} = -(x_0 - x_1)^2/4$, so that $u(a) = -u_{\min}$ in conjunction with $x_1 = a + b - x_0$ leads to the quadratic equation

$$x_0^2 - (b+a)x_0 + (b^2 + a^2 + 6ab)/8 = 0$$

Solving this equation, we find

$$x_0 = \frac{b+a}{2} \pm \frac{\sqrt{(b+a)^2 - (b^2 + a^2 + 6ab)/2}}{2}$$

whose two values are given by (7.58). Note that if we denote the smaller value by x_0, then the larger one will be precisely x_1.

Chapter 8

8.1 The algorithm may look as follows:

1. Input the value of t.
2. If $t < 0$, then replace t by $-t$.
3. Compute $m = INT[t/(2\pi)]$.
4. Replace the value of t by $t - m \cdot 2\pi$.
5. If $t > 3\pi/2$, then set $x = 2\pi - t$, $s = 1$, and go to Step 9.
6. If $t > \pi$, then set $x = t - \pi$, $s = -1$, and go to Step 9.
7. If $t > \pi/2$, then set $x = \pi - t$, $s = -1$, and go to Step 9.
8. Set $x = t$ and $s = 1$.
9. Output the values of x and s.

8.3 The area of the triangle ABC is given by $(\cos\alpha \cdot \sin\alpha)/2$ whereas the area of the triangle ADE is given by $(1\cdot\tan\alpha)/2$. On the other hand, the area of the sector ABE is seen to be given by $(\pi \cdot 1^2)\cdot \alpha/(2\pi) = \alpha/2$. Obviously,

$$\text{area } (ABC) \le \text{area } (ABE) \le \text{area } (ADE)$$

so that

$$\frac{\sin\alpha \cdot \cos\alpha}{2} \leq \frac{\alpha}{2} \leq \frac{\tan}{2}$$

or

$$\cos\alpha \leq \frac{\alpha}{\sin\alpha} \leq \frac{1}{\cos\alpha}$$

because $\sin\alpha > 0$. Now, we let the acute angle α approach zero, so that $\cos\alpha$ increases toward unity while $1/\cos\alpha$ decreases toward unity. Thus, $\alpha/\sin\alpha$, which is sandwiched between those two, also approaches unity.

8.5 Let us examine the expression $(1+1/u)^u$, where u increases indefinitely ($x = 1 + 1/u$ decreases toward unity). In Section 5.8 it was shown that the sequence $a_k = (1 + 1/k)^k$ is increasing and bounded and therefore tends to a limiting value denoted by $e \approx 2.71828$. Because the logarithmic function is monotonically increasing, we find that as x decreases toward unity the quotient $\ln x/(x-1) = \ln(1 + 1/u)^u$ is increasing (in fact, toward $\ln e = \log_e e = 1$), and thus $\ln x \leq x - 1$. The geometrical interpretation of this result can be seen by drawing (on one set of axes) the graphs of $y = \ln x$ and $y = x - 1$, in the interval $[1/e, e]$. In this interval, $y = \ln x$ increases from -1 to 1 whereas $y = x - 1$ increases from $(1/e - 1)$ to $(e - 1)$.

8.7 After "removing" the two roots from $f(x)$, we have $g(x) = f(x)/[(x-r_1)^{p_1}(x-r_2)^{p_2}]$. Suppose now that we have constructed an approximating polynomial $P_n(x)$ for $g(x)$, such that

$$|g(x) - P_n(x)| = \left|\frac{f(x)}{(x-r_1)^{p_1}(x-r_2)^{p_2}} - P_n(x)\right| < \gamma \cdot 10^{-k}$$

for all x in the relevant interval (k reflects the desired accuracy). Suppose, moreover, that we have found bounds M and m such that

$$0 < m \leq \left|\frac{1}{g(x)}\right| = \left|\frac{(x-r_1)^{p_1}(x-r_2)^{p_2}}{f(x)}\right| \leq M$$

for all x in the interval. The two preceding inequalities imply

$$\left|\frac{f(x) - (x-r_1)^{p_1}(x-r_2)^{p_2}P_n(x)}{f(x)}\right| \leq 2\gamma M \cdot \frac{1}{2} \cdot 10^{-k}$$

This is the desired inequality, analogous to (8.16).

8.8 For this case, the approximating polynomial $P_n(x)$ satisfies

$$|g(x) - P_n(x)| = |(x-s)f(x) - P_n(x)| < \gamma \cdot 10^{-k}$$

that is,

$$\left|\frac{f(x) - P_n(x)/(x-s)}{1/(x-s)}\right| < \gamma \cdot 10^{-k}$$

Suppose now that we have found bounds M and m such that

$$0 < m \leq \left|\frac{1}{g(x)}\right| = \left|\frac{1}{(x-s)f(x)}\right| \leq M$$

for all x in the relevant interval. The two preceding inequalities imply

$$\left|\frac{f(x) - \dfrac{P_n(x)}{x-s}}{f(x)}\right| < 2\gamma M \cdot \frac{1}{2} \cdot 10^{-k}$$

Note that the left-hand side is precisely the desired relative error. For a pole of order p, all we have to do is to replace $(x-s)$ by $(x-s)^p$ throughout.

8.9 Since $\tan x = \sin x / \cos x$ and since $\cos 0 = 1$, it is clear that as x approaches zero, we have

$$\frac{\tan x}{x} = \frac{1}{\cos x} \cdot \left(\frac{\sin x}{x}\right) \to 1$$

Next, to show that $\tan x$ has a simple pole at $x = \pi/2$, we must show that $1/\tan x = \cot x$ has a simple root at $x = \pi/2$. To this end, we define $x = \pi/2 - t$, so that as t approaches zero (i.e., x approaches $\pi/2$), we have

$$\frac{\cot x}{x - \pi/2} = \frac{\cot(\pi/2 - t)}{-t} = -\frac{\tan t}{t} \to -1.$$

We have thus shown that $\tan x$ is $O(x)$ as x approaches zero, and $O[1/(x-\pi/2)]$ as x approaches $\pi/2$. This shows that $\tan x$ has a simple root at $x = 0$ and a simple pole at $x = \pi/2$, so that the function $g(x)$ to be approximated by $P_n(x)$ should be $g(x) = [(x-\pi/2)/x] \tan x$. Clearly, $g(x)$ has neither a root nor a pole in the relevant interval $[0, \pi/2]$. Once $P_n(x)$ has been constructed so that it satisfies $|g(x) - P_n(x)| < \varepsilon$, the library approximation for $\tan x$ is given by $[x/(x-\pi/2)]P_n(x)$.

8.10 If we carry out the indicated multiplications in the suggested form for $P_4(x)$, we obtain

$$P_4(x) = a_4x^4 + a_4(2\alpha + 1)x^3 + a_4(\beta + \gamma + \alpha^2 + \alpha)x^2 + a_4(\alpha\gamma + \alpha\beta + \beta)x + (a_4\beta\gamma + \delta)$$

Comparing this with $P_4(x) = a_4x^4 + a_3x^3 + a_2x^2 + a_1x + a_0$, we find immediately

$$\alpha = \frac{\frac{a_3}{a_4} - 1}{2}$$

$$\beta + \gamma = \frac{a_2}{a_4} - \alpha^2 - \alpha$$

$$\beta = \frac{a_1}{a_4} - \alpha(\beta + \gamma)$$

$$\delta = a_0 - a_4\beta\gamma$$

Solving this system recursively, we obtain

$$\alpha = \frac{a_3 - a_4}{2a_4}$$

$$\beta = \frac{a_4 - a_3}{2a_4}\left(\frac{a_2}{a_4} + \frac{a_4^2 - a_3^2}{4a_4^2}\right) + \frac{a_1}{a_4}$$

$$\gamma = \frac{a_4 + a_3}{2a_4}\left(\frac{a_2}{a_4} + \frac{a_4^2 - a_3^2}{4a_4^2}\right) - \frac{a_1}{a_4}$$

$$\delta = a_0 - \beta\gamma a_4$$

in which β and γ are to be substituted from the formulas just obtained.

8.11 The requirement that the parabola $px^2 + qx + r$ must pass through the three given points implies

$$px_0^2 + qx_0 + r = y_0$$

$$px_1^2 + qx_1 + r = y_1$$

$$px_2^2 + qx_2 + r = y_2$$

We now subtract the first equation from the second and divide by $(x_1 - x_0)$, and then subtract the second equation from the third and divide by $(x_2 - x_1)$. This gives

$$p(x_1 + x_0) + q = \frac{y_1 - y_0}{x_1 - x_0}$$

$$p(x_2 + x_1) + q = \frac{y_2 - y_1}{x_2 - x_1}$$

Next, we subtract the first equation from the second and divide by $(x_2 - x_0)$ to obtain

$$p = \frac{\dfrac{y_2 - y_1}{x_2 - x_1} - \dfrac{y_1 - y_0}{x_1 - x_0}}{x_2 - x_0}$$

The first of the previous pair of equations immediately gives

$$q = \frac{y_1 - y_0}{x_1 - x_0} - p(x_0 + x_1)$$

into which the formula just obtained for p should be substituted. Finally, the expression for r in (8.27) can be obtained by substituting the formula for q just obtained in the first of the triplet of equations above. Next, rearrange the expression for p just obtained as a sum of three fractions with numerators y_0, y_1, and y_2, respectively. You will thus get

$$p = \frac{y_0}{(x_0 - x_1)(x_0 - x_2)} + \frac{y_1}{(x_1 - x_0)(x_1 - x_2)} + \frac{y_2}{(x_2 - x_0)(x_2 - x_1)}$$

which is precisely the coefficient of x^2 in both the Lagrangian and Newton's forms, as detailed in the solution of Exercise 7.10. Moreover, it can be shown analogously that the coefficient of x and the constant term (i.e., q and r) coincide, respectively, with the values $B = \widetilde{B}$ and $C = \widetilde{C}$ as given in the solution of Exercise 7.10. These results are not surprising in view of the reasoning contained in the solution of Exercise 7.2.

8.16 The polynomial $P_{m+n}(x)$ is of the form

$$P_{m+n}(x) = a_{m+n}x^{m+n} + a_{m+n-1}x^{m+n-1} + \cdots + a_1 x + a_0$$

and clearly possesses $k = m + n + 1$ coefficients – that is, k free parameters to be determined so as to pass through k arbitrary distinct interpolation points. The rational function $R_{m,n}(x)$ is of the form

$$R_{m,n}(x) = \frac{a_m x^m + a_{m-1}x^{m-1} + \cdots + a_1 x + a_0}{b_n x^n + b_{n-1}x^{n-1} + \cdots + b_1 x + b_0}$$

but since $b_n \neq 0$, due to the assumption that the denominator is genuinely of degree n, we can divide both numerator and denominator by b_n. This casts $R_{m,n}(x)$ into the form

$$R_{m,n}(x) = \frac{A_m x^m + A_{m-1} x^{m-1} + \cdots + A_1 x + A_0}{x^n + B_{n-1} x^{n-1} + \cdots + B_1 x + B_0}$$

which possesses $k = m + n + 1$ coefficients (i.e., free parameters). This explains why the number of free parameters is again k and not $(k+1)$. An example of the use of these parameters in conjunction with three interpolation points, for the case $m = n = 1$, is shown in (8.32) in the text.

Index